A CHALLENGE OF NUMBERS

People in the Mathematical Sciences

Prepared by
Bernard L. Madison
and
Therese A. Hart
for the
Committee on the Mathematical Sciences
in the Year 2000

Mathematical Sciences Education Board
Board on Mathematical Sciences
Commission on Physical Sciences, Mathematics, and Resources
National Research Council

National Academy Press
Washington, D.C. 1990

NOTICE: The project that is the subject of this report was approved by the Governing Board of the National Research Council, whose members are drawn from the councils of the National Academy of Sciences, the National Academy of Engineering, and the Institute of Medicine. The members of the committee responsible for the report were chosen for their special competences and with regard for appropriate balance.

This report has been reviewed by a group other than the authors according to procedures approved by a Report Review Committee consisting of members of the National Academy of Sciences, the National Academy of Engineering, and the Institute of Medicine.

The National Academy of Sciences is a private, nonprofit, self-perpetuating society of distinguished scholars engaged in scientific and engineering research, dedicated to the furtherance of science and technology and to their use for the general welfare. Upon the authority of the charter granted to it by the Congress in 1863, the Academy has a mandate that requires it to advise the federal government on scientific and technical matters. Dr. Frank Press is president of the National Academy of Sciences.

The National Academy of Engineering was established in 1964, under the charter of the National Academy of Sciences, as a parallel organization of outstanding engineers. It is autonomous in its administration and in the selection of its members, sharing with the National Academy of Sciences the responsibility for advising the federal government. The National Academy of Engineering also sponsors engineering programs aimed at meeting national needs, encourages education and research, and recognizes the superior achievements of engineers. Dr. Robert M. White is president of the National Academy of Engineering.

The Institute of Medicine was established in 1970 by the National Academy of Sciences to secure the services of eminent members of appropriate professions in the examination of policy matters pertaining to the health of the public. The Institute acts under the responsibility given to the National Academy of Sciences by its congressional charter to be an adviser to the federal government and, upon its own initiative, to identify issues of medical care, research, and education. Dr. Samuel O. Thier is president of the Institute of Medicine.

The National Research Council was organized by the National Academy of Sciences in 1916 to associate the broad community of science and technology with the Academy's purposes of furthering knowledge and advising the federal government. Functioning in accordance with general policies determined by the Academy, the Council has become the principal operating agency of both the National Academy of Sciences and the National Academy of Engineering in providing services to the government, the public, and the scientific and engineering communities. The Council is administered jointly by both Academies and the Institute of Medicine. Dr. Frank Press and Dr. Robert M. White are chairman and vice chairman, respectively, of the National Research Council.

The Committee on the Mathematical Sciences in the Year 2000, which was appointed at the beginning of 1988, is a three-year project of the Mathematical Sciences Education Board and the Board on Mathematical Sciences. Its purpose is to provide a national agenda for revitalizing mathematical sciences education in U.S. colleges and universities.

Support for this project and for the publication and dissemination of this report was provided by grants from the National Science Foundation and the National Security Agency.

Cover photograph reprinted courtesy of the University of Maryland and with permission from John Consoli, photographer. Copyright © 1989 by John Consoli.

Library of Congress Catalogue Card Number 89-64078

International Standard Book Number 0-309-04190-2

Available from:
National Academy Press
2101 Constitution Avenue, N.W.
Washington, D.C. 20418

SO89

Summaries of this report may be obtained from MS 2000, 2101 Constitution Avenue, N.W., Washington, D.C. 20418

Printed in the United States of America

Committee on the Mathematical Sciences in the Year 2000

William E. Kirwan (Chairman), President, University of Maryland
Ramesh A. Gangolli (Vice Chairman), Professor of Mathematics, University of Washington
Lida K. Barrett, Dean, College of Arts & Sciences, Mississippi State University
Maria A. Berriozabal, Councilwoman, City of San Antonio, Texas
Ernest L. Boyer, President, Carnegie Foundation for the Advancement of Teaching
William Browder, Professor of Mathematics, Princeton University
Rita R. Colwell, Director of the Maryland Biotechnology Institute and Professor of Microbiology, University of Maryland
John M. Deutch, Provost, Massachusetts Institute of Technology
Ronald G. Douglas, Dean, Division of Physical Sciences and Mathematics, State University of New York, Stony Brook
Patricia A. Dyer, Vice President for Academic Affairs, Palm Beach Community College
Lloyd C. Elam, Distinguished Service Professor of Psychiatry, Meharry Medical College
Sheldon L. Glashow, Higgins Professor of Physics, Harvard University
Nancy J. Kopell, Professor of Mathematics, Boston University
Donald W. Marquardt, Consultant Manager, E. I. du Pont de Nemours & Co.
David S. Moore, Professor of Statistics, Purdue University
Jaime Oaxaca, Vice Chairman, Coronado Communications
Moshe F. Rubinstein, Professor of Engineering and Applied Science, University of California, Los Angeles
Ivar Stakgold, Chairman, Department of Mathematical Sciences, University of Delaware
S. Frederick Starr, President, Oberlin College
Lynn Arthur Steen, Professor of Mathematics, St. Olaf College

Staff

James A. Voytuk, Project Director
Bernard L. Madison, Project Director (through August 1988)
Therese A. Hart, Research Associate (through November 1989)
Craig E. Hicks, Project Assistant

Mathematical Sciences Education Board

Alvin W. Trivelpiece (Chairman), Director, Oak Ridge National Laboratory

Shirley A. Hill (Past Chairman), Curators' Professor of Mathematics and Education, University of Missouri—Kansas City

Iris M. Carl (Vice Chairman), Elementary Mathematics Instructional Supervisor, Houston Independent School District, Texas

Lillian C. Barna, Superintendent of Schools, Tacoma Public Schools, Washington

Lida K. Barrett, Dean, College of Arts and Sciences, Mississippi State University

C. Diane Bishop, Superintendent of Public Instruction, State of Arizona

Constance Clayton, Superintendent of Schools, School District of Philadelphia, Pennsylvania

Paula B. Duckett, Elementary Mathematics Teacher, River Terrace Community School, Washington, D.C.

Joan Duea, Elementary School Teacher, Price Laboratory School and Professor of Education, University of Northern Iowa

Joseph W. Duncan, Corporate Vice President and Chief Economist, The Dun & Bradstreet Corporation

Wade Ellis, Jr., Mathematics Instructor, West Valley College, California

Shirley M. Frye, Director of Curriculum and Instruction, Scottsdale School District, Arizona

Ramanathan Gnanadesikan, Head, Information Science Research Division, Bell Communications Research

Donald L. Kreider, Vice Chairman, Mathematics and Computer Science Department, Dartmouth College

Martin D. Kruskal, Professor of Mathematics, Rutgers University

Katherine P. Layton, Mathematics Teacher, Beverly Hills High School, California

Steven J. Leinwand, Mathematics Consultant, Connecticut State Department of Education

Richard S. Lindzien, Sloan Professor of Meteorology, Massachusetts Institute of Technology

Gail Lowe, Principal, Acacia Elementary School, Thousand Oaks, California

Steven P. Meiring, Mathematics Specialist, Ohio State Department of Education

Jose P. Mestre, Associate Professor of Physics, University of Massachusetts

Calvin C. Moore, Associate Vice President, Academic Affairs, University of California, Berkeley

Jo Ann Mosier, Mathematics Teacher, Fairdale High School, Louisville, Kentucky

Leslie Hiles Paoletti, Chairman, Department of Mathematics and Computer Science, Choate Rosemary Hall, Connecticut

Lauren B. Resnick, Director, Learning Research and Development Center, University of Pittsburgh; liaison with the Commission on Behavioral and Social Sciences and Education, National Research Council

Yolanda Rodriguez, Middle School Mathematics Teacher, Martin Luther King School, Cambridge, Massachusetts

Thomas A. Romberg, Professor of Curriculum and Instruction, University of Wisconsin, Madison

Isadore M. Singer, Institute Professor, Department of Mathematics, Massachusetts Institute of Technology

Lynn Arthur Steen, Professor of Mathematics, St. Olaf College

William P. Thurston, Professor of Mathematics, Princeton University

Manya S. Ungar, Past President, The National Congress of Parents and Teachers

Zalman Usiskin, Professor of Education, The University of Chicago

John B. Walsh, Vice President/Chief Scientist, Boeing Military Airplanes

Nellie C. Weil, Past President, National School Boards Association

Guido L. Weiss, Elinor Anheuser Professor of Mathematics, Washington University; liaison with the Commission on Physical Sciences, Mathematics, and Resources, National Research Council

Staff

Kenneth M. Hoffman, Executive Director

Marcia P. Sward, Executive Director (through August 1989)

Board on Mathematical Sciences

Phillip A. Griffiths (Chairman), Provost and James B. Duke Professor of Mathematics, Duke University
Lawrence D. Brown, Professor of Mathematics, Cornell University
Ronald G. Douglas, Dean, College of Physical Sciences and Mathematics, State University of New York, Stony Brook
David Eddy, J. Alexander McMahon Professor of Health Policy and Management, Duke University
Frederick W. Gehring, Professor of Mathematics, University of Michigan
James G. Glimm, Chairman, Department of Applied Mathematics and Statistics, State University of New York, Stony Brook
William H. Jaco, Executive Director, American Mathematical Society
Joseph Kadane, Professor of Statistics, Carnegie-Mellon University
Gerald J. Lieberman, Professor of Operations Research and Statistics, Stanford University
Alan Newell, Head, Department of Mathematics, University of Arizona
Jerome Sacks, Head, Department of Statistics, University of Illinois
Guido L. Weiss, Elinor Anheuser Professor of Mathematics, Washington University
Shmuel Winograd, Director, Mathematical Sciences Department, IBM Thomas J. Watson Research Center

Staff

Lawrence H. Cox, Director

Commission on Physical Sciences, Mathematics, and Resources

Norman Hackerman (Chairman), Chairman, Scientific Advisory Board, Robert A. Welch Foundation
Robert C. Beardsley, Senior Scientist and Chairman, Department of Physical Oceanography, Woods Hole Oceanographic Institution
B. Clark Burchfiel, Schlumberger Professor of Geology, Massachusetts Institute of Technology
George F. Carrier, Professor Emeritus, Harvard University
Ralph J. Cicerone, Chair, Geosciences Department, University of California at Irvine
Herbert D. Doan, The Dow Chemical Company (retired)
Peter S. Eagleson, Edmund K. Turner Professor of Civil Engineering, Massachusetts Institute of Technology
Dean E. Eastman, Vice President, IBM T. J. Watson Research Center
Marye Anne Fox, Roland Pettit Centennial Professor of Chemistry, University of Texas
Gerhart Friedlander, Consultant, Brookhaven National Laboratory
Lawrence W. Funkhouser, Chevron Corporation (retired)
Phillip A. Griffiths, Provost and James B. Duke Professor of Mathematics, Duke University
Neal F. Lane, Provost, Rice University
Christopher F. McKee, Professor of Physics and of Astronomy, University of California at Berkeley
Richard S. Nicholson, Executive Director, American Association for the Advancement of Science
Jack E. Oliver, Director of INSTOC, Cornell University
Jeremiah P. Ostriker, Chairman, Department of Astrophysical Sciences, Director, Princeton University Observatory, Princeton University
Philip A. Palmer, Principal Consultant, E. I. du Pont de Nemours & Company
Frank L. Parker, Professor of Civil and Environmental Engineering, Vanderbilt University
Denis J. Prager, Deputy Director, Health Program, MacArthur Foundation
David M. Raup, Professor, Department of Geophysical Sciences, University of Chicago
Roy F. Schwitters, Director, Superconducting Super Collider Laboratory
Larry L. Smarr, Director of National Center for Supercomputing Applications, Professor of Astronomy and Physics, University of Illinois at Urbana-Champaign
Karl K. Turekian, Silliman Professor of Geology and Geophysics, Yale University

Staff

Myron F. Uman, Acting Executive Director

Preface

During the last decade of this century, the U.S. educational system will face many challenges, but few more important than the renewal and revitalization of the mathematical sciences. Problems in mathematical sciences education exist at all levels—elementary, high school, and postsecondary. These problems are caused by many factors, including a static curriculum, insufficient resources devoted to instruction, inadequate numbers of well-trained teachers, declining student interest, and more generally, the public's failure to understand fully the importance of education in science for the well-being of our society. The problems manifest themselves in the product of the educational enterprise—the scientifically trained work force that must serve the needs of our society. The demands of the next decade and the twenty-first century will require that large numbers of America's work force be equipped with higher levels of mathematical sophistication. Yet, the National Science Foundation and other organizations project that by the year 2000, unless there is a dramatic change, our nation will face a significant shortfall of people with the necessary skills in the mathematical sciences.

In response to this crucial situation the National Research Council has established the Committee on the Mathematical Sciences in the Year 2000, under the direction of the Mathematical Sciences Education Board and the Board on Mathematical Sciences, to assess the present state of collegiate education in the mathematical sciences, to identify existing problems, and to recommend action that can remedy this situation. It is hoped that this analysis by the committee will lead to a national agenda for renewal and revitalization of the mathematical sciences and to a strategy for implementation that will stimulate all sectors of our society into action.

The present descriptive report, *A Challenge of Numbers*, is one of three planned by the committee. It describes the circumstances and issues related to the human resources in the mathematical sciences, principally students and teachers at U.S. colleges and universities. The mathematical sciences, based largely in academia, are crucial not only for the development of new knowledge but also as a needed resource in the education of our technological work force. As with other scientific professions, the challenge of meeting the growing need for workers in the face of shrinking supplies will require a program of national action, but the situation in the mathematical sciences is particularly severe. Attrition from the educational pipeline in mathematics has given rise to low degree production at all levels in colleges and universities, and socioeconomic forces have resulted in unprecedentedly low interest in mathematics as a major among entering freshmen. At the same time, growing enrollments in lower-level mathematics courses with small or no increases in staffing levels have created less than an ideal environment for faculty in the mathematical sciences. Because of the fundamental role of mathematics in support of science and engineering, the problems in the U.S. educational system will not be corrected unless the problems in mathematics education are corrected. On the other hand, a revitalization of the mathematical sciences could lead the way in effecting change across the educational spectrum.

A Challenge of Numbers contains data from a large number of existing sources. It provides in one place a comprehensive set of data describing the demographic situation in the mathematical sciences. This report does not provide answers, nor does it draw conclusions. Rather, it serves as a reference. As such, it will be an important source of data for the Committee on the Mathematical Sciences in the Year 2000 in formulating its recommendations. Also, it is hoped that the report will be used by educators and administrators in industry, government, and education to alert policymakers about the urgent need for revitalization of the mathematical sciences.

This is the second report of the MS 2000 project. The first, *Everybody Counts*, was published jointly with the Mathematical Sciences Education Board and the Board on Mathematical Sciences and provided a general analysis of the trends and needs in mathematics education. The next review paper, to be published in late spring 1990, will assess the college and university curriculum, and the final paper, due in summer 1990, will report on resources available for instruction in the mathematical sciences. Following the completion of the review papers, the Committee on the Mathematical Sciences in the Year 2000 will be in a position to issue its final recommendations in a report to the nation scheduled for late 1990.

On behalf of the committee, I want to express our gratitude to Bernard L. Madison and Therese A. Hart for their work in compiling the necessary background information and for writing *A Challenge of Numbers*. They have provided an important service to the members of the Committee on the Mathematical Sciences in the Year 2000 and to the mathematical sciences community by making these data available in a convenient and accessible format.

William E. Kirwan
Chairman, Committee on the Mathematical Sciences
 in the Year 2000
President, University of Maryland at College Park

Contents

Preface viii

List of Figures xii

List of Tables xiv

List of Boxes xv

1 Introduction and Historical Perspective   1

The MS 2000 Project and the Scope of This Report 1
Three Roller Coaster Decades 2
National Efforts Toward Renewal 7
Contents of This Report 8

2 The U.S. Labor Force and Higher Education   9

Introduction 9
More Skills and Greater Adaptability 10
Growth in Science-Based Occupations 11
Higher Education in the United States 12
The Pool of Potential Students and Workers 13
Persistence in College Enrollment 14
Shifting Interests of College Students 15
Natural Sciences and Engineering 16
The Challenges and the Responsibility 18

3 College and University Mathematical Sciences   19

Introduction 19
Strong at the Top 20
Mixed Precollege Indicators 22
Troublesome Transitions from High School to College 24
Remediation in College 28
Service Courses 30
Mathematics as an Academic Competency and Subject 32

4 Majors in Mathematics and Statistics — 35

Introduction 35
Undergraduate Majors 37
Degrees for Secondary School Mathematics Teachers 40
Graduate Students 42
Master's Degree Recipients 45
Doctoral Degree Recipients 48
Patterns and Prospects 52

5 Mathematical Scientists in the Workplace — 53

Introduction 53
General Characteristics and Trends 54
Employment of Recent Graduates 56
Secondary School Mathematics Faculty 58
Characteristics of College and University Faculties 61
What Faculty Members Do 62
Faculty Members by Duties and Credentials 65
The Research Faculty 67
Faculty Salaries 68
Ages of Faculty Members 68
Women and Minorities on the Faculty 70
Two-Year College Faculty Mobility 70
Four-Year College and University Doctorate Faculty 70
Four-Year College and University Nondoctorate Faculty 70
Summary 71

6 Issues and Implications — 73

Bibliography — 75

Appendix Tables — 87

List of Figures

1.1	Total number of scientists and engineers.	2
1.2	Ph.D. degrees in mathematics, 1986-1987.	3
1.3	Left: Total undergraduate enrollments in mathematical sciences departments. Right: Mathematical sciences faculty at colleges and universities.	4
1.4	Mathematical sciences degrees awarded.	6
2.1	The educational requirements of the work force are increasing.	10
2.2	Percent distribution of undergraduate enrollments by race and ethnic group.	13
2.3	The pool of college students is changing, 18- to 24-year-old population.	13
2.4	Percent of 18- and 19-year-olds who are high school dropouts, by ethnic group.	14
2.5	Enrollment in institutions of higher education as a percent of high school graduates.	14
2.6	Shifting interest in selected majors.	17
3.1	SAT mathematics scores, 1967 to 1987.	22
3.2	ACT mathematics scores, 1973 to 1988.	22
3.3	Percent increase in enrollments in selected mathematics courses in colleges and universities, 1965 to 1985.	30
3.4	Undergraduate enrollments in mathematical sciences departments at U.S. colleges and universities.	32
4.1	Students in the mathematical sciences pipeline—about half are lost each year.	36
4.2	A representation of U.S. students in the mathematics pipeline.	36
4.3	Number of mathematical sciences degrees awarded by U.S. institutions, 1950 to 1986.	37
4.4	Percentage of entering college freshmen expecting to major in mathematics.	38
4.5	Bachelor's degrees awarded in the mathematical sciences, 1970 to 1986.	38
4.6	Number of bachelor's degrees awarded in mathematical and computer sciences, 1970 to 1986.	39
4.7	Expected versus actual number of bachelor's degrees in mathematical sciences.	41
4.8	Left: Interest in mathematics and education among entering college freshmen. Right: Degrees in mathematics and education among exiting college seniors.	43
4.9	Percent of full-time graduate students in doctorate-granting institutions who are non-U.S. citizens, 1975 and 1986.	44
4.10	Mathematical sciences graduate students enrolled full-time in doctorate-granting institutions, 1975 to 1986.	44

4.11	Percent of non-U.S. citizens as mathematical sciences graduate students by type of institution, 1977 to 1986.	44
4.12	Source of major support for mathematical sciences graduate students in doctorate-granting institutions, 1986.	45
4.13	Types of major support for graduate students in doctorate-granting institutions, 1986.	46
4.14	Master's degrees awarded, mathematical sciences.	46
4.15	Master's degrees in mathematical sciences, distribution by subfield.	47
4.16	Ph.D. degrees in mathematics.	48
4.17	Doctoral degrees in mathematical sciences, distribution by subfield and sex.	51
4.18	Number of doctorate recipients in broadly interpreted mathematical sciences.	51
5.1	Percent of recent mathematics degree holders employed in a science or engineering job, 1976 and 1986.	55
5.2	Field of employment for recent mathematics degree recipients, 1986.	57
5.3	Primary work activities of recent mathematics degree recipients, 1986.	58
5.4	Median annual salaries of recent science and engineering graduates.	58
5.5	Supply and demand of new elementary and secondary school teachers, 1970 to 1992.	59
5.6	Left: Number of full-time mathematical sciences faculty members at colleges and universities. Right: Number of part-time mathematical sciences faculty members at colleges and universities.	60
5.7	Mathematical and computer sciences enrollments per FTE of faculty.	61
5.8	Mathematical sciences faculty salaries, 1970 to 1985 (in 1985 dollars).	63
5.9	Age distribution of full-time mathematical sciences faculty in four-year colleges and universities.	64
5.10	Estimated number of retirements of full-time college and university mathematical sciences faculty.	65
5.11	Top: Source of new hires of two-year college full-time faculty in mathematical sciences. Bottom: Destination of departing mathematical sciences two-year college full-time faculty.	66
5.12	Source of two-year college part-time faculty in mathematical sciences.	68
5.13	Left: Source of new hires of doctorate faculty in mathematical sciences, 1983 to 1988. Right: Destination of departing mathematical sciences doctoral faculty, 1983 to 1988.	69

List of Tables

2.1	Attainment rates of advanced degrees for selected fields, 1971 to 1985	18
4.1	Mathematical sciences bachelor's degrees per 1,000 mathematical sciences enrollments, 1965 to 1985	37
4.2	Changes in mathematical sciences majors by undergraduate grade point averages, 1981 freshman cohort	39
4.3	1985 bachelor's degrees awarded in mathematical sciences	40
4.4	Summary of responses on quality and quantity of undergraduate majors	41
4.5	1987 SAT scores by intended college major	42
4.6	1986 GRE scores by undergraduate and intended graduate major	42
4.7	Summary of responses on quality and quantity of graduate students	45
4.8	1985 master's degrees awarded in mathematical sciences programs	47
4.9	Attainment rates of master's and doctoral degrees	48
4.10	Ratio of new doctorates in mathematics to new doctorates in selected other fields, 1970 to 1985	49
4.11	Mathematics majors going on to doctoral study in other areas of science and engineering, l960 to 1985	49
4.12	Ethnic representations among new mathematical sciences doctorates, U.S. citizens, 1977 to 1986	50
4.13	Ethnic representation among all new research doctorates, U.S. citizens and permanent residents, 1977 to 1986	50
5.1	Estimates of the number of mathematical scientists by National Science Foundation (NSF), Bureau of Labor Statistics (BLS), and Conference Board for Mathematical Sciences (CBMS)	54
5.2	Type of employer of mathematical scientists by degree level, 1986	56
5.3	Professional activities of four-year college and university mathematical sciences faculty	62
5.4	Professional activities of two-year college mathematical sciences faculty	64
5.5	Numbers of mathematical sciences faculty members by teaching area and type of institution, 1987	67
5.6	Age distribution of mathematical sciences faculty members in 105 research universities, 1980 and 1986	68
5.7	Full-time mathematical sciences faculty by ethnic origin and sex, 1985	69
5.8	Estimate of average annual net flow into doctoral faculty at four-year colleges and universities, 1982 to 1987	71

List of Boxes

1.1	Computer Science	5
1.2	Sources of Data	7
1.3	Statistics	8
2.1	Degree Programs in Mathematics	11
2.2	Degree Programs in Statistics	12
3.1	Professional Organizations	21
3.2	AMS-MAA Survey Reports	25
3.3	CBMS Surveys	26
3.4	Minorities and Women	27
3.5	Intervention Programs	28
3.6	The Texas Prefreshman Engineering Program	29
3.7	Professional Development Program	31
3.8	The Mathematics, Engineering, Science Achievement Program	33

1 Introduction and Historical Perspective

Mathematics has become essential and pervasive in the U.S. workplace, and projections indicate that its use will expand, as will the need for more workers with a knowledge of college-level mathematics. However, socioeconomic and demographic projections as well as circumstances within the college and university mathematical sciences[1] system suggest that an adequate supply of appropriately educated workers is not forthcoming. Development of mathematical talent will be impeded by the low general interest in mathematics as a college major; the relatively small numbers of minorities and women studying and practicing mathematics; a shortage of qualified faculty to deal with huge enrollments in low-level courses and students with widely varying levels of preparation; and the difficulty of maintaining the vitality of the mathematical sciences faculty.

The MS 2000 Project and the Scope of This Report

Because a healthy flow of mathematical talent is important for the nation's welfare, the National Research Council initiated in 1986 the project Mathematical Sciences in the Year 2000 (MS 2000) to assess the status of college and university mathematical sciences and to design a plan for revitalization and renewal. This report describes the circumstances and issues surrounding the people involved in the mathematical sciences, principally students and teachers. The description is not complete because comprehensive data are not available, but most data that are relevant and available are included and are adequate to describe the circumstances in the mathematical sciences. Two additional descriptive reports—one on curriculum and the other on resources—are forthcoming. Together these three reports will form the basis for the the MS 2000 Committee's final report, which will contain recommendations for actions to achieve revitalization and renewal of the college and university mathematical sciences enterprise.

This report is concerned with all students of collegiate mathematics. However, mathematics majors have a special role to play because they are the source of the new faculty members necessary to renew and sustain the system. And increases in the need for mathematics in the workplace in turn fuel a need for more academically skilled workers. A dramatic demonstration of this need is the

---

[1] For the purposes of this report the discipline referred to as the "mathematical sciences" includes mathematics, applied mathematics, and statistics. A broader definition is generally used in the taxonomy of scientific disciplines. For a discussion of the mathematical sciences research community see *Renewing U.S. Mathematics: Critical Resource for the Future* (National Academy Press, Washington, D.C., 1984), pp. 77-85. Computer science is not a branch of the mathematical sciences, but its close ties with mathematics, both intellectually and administratively, have significantly affected college and university mathematical sciences over the past two decades. This report does not attempt to describe circumstances in computer science, but references to computer science are necessary because of these ties and their effects.

doubling of the number of scientists and engineers in a single decade (Figure 1.1).

Understanding students and teachers in the mathematical sciences—who they are, what they learn and teach, and how they use what they learn—requires understanding the vast and diverse system in which they work. The mathematical sciences programs in U.S. colleges and universities account for nearly 10% of all collegiate teaching in the United States and nearly 30% of all collegiate teaching in the natural sciences and engineering. Each term, approximately 3 million students are taught by more than 40,000 full-time and part-time faculty members and 8,000 graduate teaching assistants in 2,500 institutions. To better understand this system and how current circumstances evolved, a review of events of the past 30 years is helpful.

Three Roller Coaster Decades

For centuries, mathematics has been recognized as interesting, challenging, and essential for the support of science and engineering. Within this century, mathematics has become much more broadly applicable and important. Giant strides toward recognition of its significance were made during World War II. After World War II, U.S. mathematicians branched out, studying and developing new areas in many directions very successfully. This period of innovation and the concurrent expansion of college and university mathematics programs positioned mathematics as a key participant in the nation's emphasis on science spurred on by the 1957 launch of Sputnik. Thus began three decades of extraordinary change—a decade of expansion, followed by a decade of adjustment and depression, followed by a decade of partial recovery.

The decade following Sputnik's launch was one of expansion for U.S. mathematics. Statistics became more widely recognized as a distinct discipline and began to flourish. Then—as now, to a slightly lesser extent—most of the research in mathematics and statistics was performed in universities. College enrollments increased, faculties expanded, and positions were plentiful. The number of bachelor's degrees in mathematics awarded annually tripled, and the number of graduate degrees increased fivefold in this decade. Support for specialized research programs, which was available from federal agencies for individuals, was ideally suited to the mathematical research mode.

In the late 1960s, immediately following the dramatic expansion of science and mathematics programs, the nation's interest and attention shifted to social issues. Although more students continued to enter college as access to higher education expanded significantly, many came without adequate preparation for college mathematics and with questions about the relevance of learning any. Over the 20-year period from 1965 to 1985, college enrollments doubled, and mathematical sciences enrollments more than kept pace. However, most of the increase in mathematics enrollments was at the lower levels, with remedial enrollments in high school mathematics taught in college leading the way.

The surge in the numbers of degrees awarded in the mathematical sciences in the late 1960s and early 1970s and the lack of established employment markets for mathematicians outside of academe created more degree holders than there were jobs, especially at the doctoral level; in addition, part of the response to increased enrollments in mathematics courses was to let student-faculty ratios increase. A depressed employment market resulted that

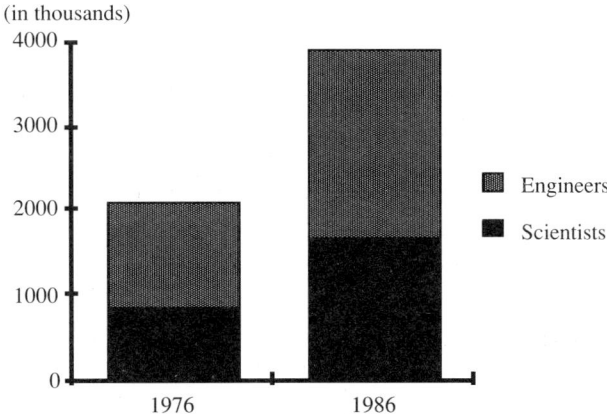

FIGURE 1.1 Total number of scientists and engineers. SOURCE: National Science Board (NSB, 1987).

lasted nearly a decade, into the early 1980s. To some extent this depression was spread across all science and engineering fields. Statistics was an exception, with some modest increases in degrees granted and a better nonacademic employment market.

College and university mathematical sciences faculties were changing. Increasing responsibilities for teaching precalculus and high-school-level courses, the prediction that college enrollments would soon decline, and the perception that mathematics Ph.D.s were plentiful changed employment practices on college faculties. The changes included the creation of positions with heavy teaching loads for full-time faculty and the use of more part-time and temporary teachers. Many faculty members had little time and motivation for personal scholarship; some lapsed into inactivity. Teaching introductory algebra and calculus to students majoring in other areas became more widespread and restricted the independent growth of mathematics and mathematicians. Some faculty members did not teach what they thought about—their research—and also had little enthusiasm or latitude to think about what they taught. These forces reduced the attention to curriculum development and reform. In response to nationally articulated goals in the mid-1960s, the fraction of Ph.D.s on mathematical sciences faculties had increased significantly to nearly 80% in four-year institutions, but a seeming mismatch between training and duties prompted a reversal of this effort. In particular, new doctoral degree holders, educated for research, were mismatched with the teaching positions available. Consequently, both teaching and research suffered.

In research universities, graduate students, plentiful in the 1960s, assumed a large share of the teaching responsibilities. Information on a weak mathematics employment market and better opportunities in other areas such as computer science spread quickly among U.S. students, and the numbers choosing mathematics as a major area of study began to decline. This decline was partially offset at the graduate level by increases in the number of non-U.S. students that, combined with the significant decline in the number of U.S. students enrolled in mathematics, changed the non-U.S. representation from about one of five in 1970

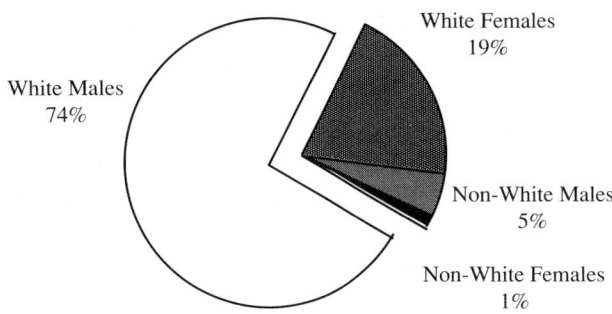

FIGURE 1.2 Ph.D. degrees in mathematics, 1986-1987. SOURCE: American Mathematical Society (AMS, 1987).

to nearly one of two in 1988. This trend, coupled with the heavy teaching burden carried by graduate students, created teaching problems across the country.

Factors other than the poor employment market also reduced the number of mathematical sciences majors. One factor was the predominance of white middle-class males in the study and practice of mathematics. Relatively few women and minorities were choosing mathematically based careers and curricula, although more women, more blacks, and more Hispanics were entering college. The fraction of bachelor's degrees earned by women did increase—from about one-third of the total in the mid-1960s to almost one-half by 1986—but the increase was smaller at the master's level and smaller still at the doctoral level (Figure 1.2). That comparatively few blacks and Hispanics choose mathematically based careers has continued to be the case. The number of Native Americans choosing such careers is small but does reflect approximately this group's share of the total U.S. population, while Asian-Americans continue to show a preference for these careers.

In the 1960s and 1970s, little attention was given to creating employment opportunities for mathematical scientists in the nonacademic workplace. Academic employment in a research environment was the dominant destination for degree holders in mathematics, and these opportunities had diminished. Thus, not only were there rough transitions to the workplace for those with bachelor's and master's degrees, but Ph.D.s, too, were not suitably matched with the teaching jobs that were available in colleges.

Mathematical sciences enrollments in introductory and remedial courses continued to increase in the 1970s, fueled by added mathematics requirements in the curricula of fields such as business and by shifts to majors that required more mathematics. This development reflected an increasing need for mathematics in the workplace, both for professionals in other areas and for mathematical scientists. Mathematics was emerging as more important in professional education, achieving a new prominence that complemented its centuries-long role in human intellectual development. Problem-solving ability and adaptability dominated the requirements of new jobs. Said another way, liberal arts education—especially mathematics education—was becoming closer to professional education. However, the nonacademic employment market for mathematical scientists continued to be poorly understood and was invisible to many.

Departments across the country met the increased enrollments of the 1970s with a variety of types of faculty members and the same traditional courses, mostly because they were busy and lacked resources (Figure 1.3). Many temporary and part-time teachers were hired on an ad hoc basis term after term. Thus began a dismantling of the buildup to a high fraction of faculty with Ph.D.s that had just been achieved. The responsibilities of departments became more diverse and more difficult to carry out because of heavier involvement in coordinating activities, fewer experienced and involved teachers, large remedial and placement problems, and fewer mathematics majors. No other collegiate discipline teaches as many students with such widely differing levels of preparation as does mathematics, and most of the students are expected to use the mathematics in subsequent courses. An overwhelming combination of problems of collegiate teaching—unmotivated and underprepared students, unenthusiastic teachers, language problems in the classroom, outdated and irrelevant curricula and courses, large classes, heavy teaching loads, too few resources, and little use of modern technology—came together in the 1970s and resonated in mathematics classrooms across the United States.

By the early 1980s, the number of degrees awarded annually in the mathematical sciences had fallen by nearly 50% at all three levels (Figure 1.4). Occurring simultaneously with the decline in the numbers of degrees awarded were increases in enrollments and in reliance on part-time faculty. Belatedly, institutions decided that the high mathematical sciences enrollments would persist, and they began to employ regular faculty members. By this time there were too few U.S. citizens among the new doctoral degree holders to meet the demand; indeed, there were overall shortages of candidates. There were (and are) no surplus pools of mathematical sciences Ph.D.s—no large number

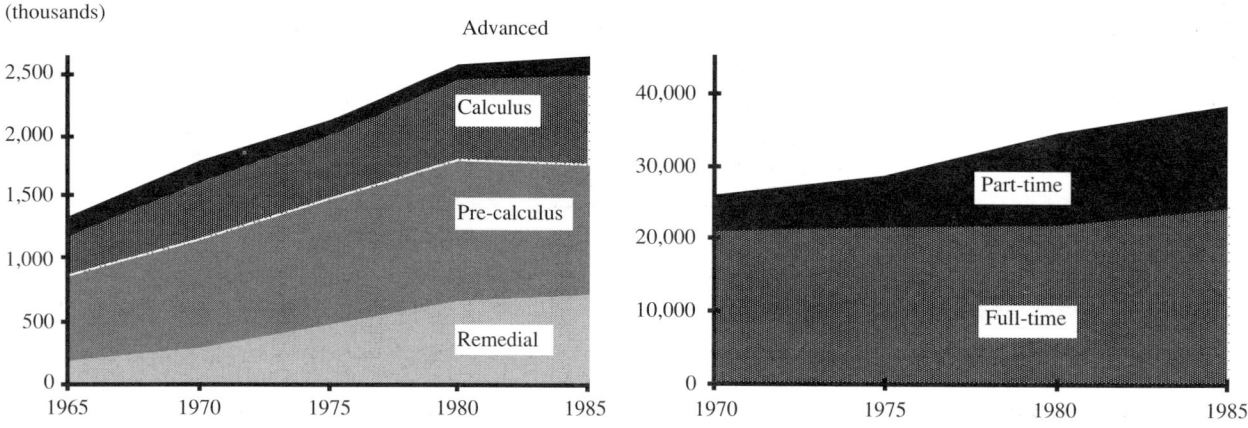

FIGURE 1.3 Left: Total undergraduate enrollments in mathematical sciences departments. Right: Mathematical sciences faculty at colleges and universities. SOURCE: Conference Board of the Mathematical Sciences (CBMS, 1987).

BOX 1.1 Computer Science

Computer science has developed since World War II from roots in mathematics and electrical engineering. It has become a separate academic discipline within the past two decades and has developed its own sources of students and federal funding for research. (See the 1984 David Report (NRC, 1984) for a fuller analysis.) Computer science is not considered to be a part of the mathematical sciences, and that is the position taken throughout this report. However, older reports on the mathematical sciences may include computer science data, and computer science and mathematics continue to be administered by the same unit in most colleges and universities (see Box 3.3). Because of these historical, administrative, and intellectual ties, the emergence and rapid growth of the discipline of computer science have had a significant effect on the mathematical sciences. At points in this report some of these effects are conjectured, but only the effects of computer science on the mathematical sciences are considered. No attempt is made to describe the conditions in computer science. Nevertheless, the grouping of computer science with the mathematical sciences in many college and university departments and the dual teaching roles of many faculty members are facts.

The 1985 CBMS Survey (Box 3.3) concluded that in four-year colleges and universities, half (49%) of all computer science course sections were taught in departments with mathematics and the other half (51%) in computer science departments. (This breakdown did not include courses in computing taught by many units in business and engineering, for example.) Thus approximately 270,000 students enrolled in computer science courses were taught in mathematics departments in fall 1985. This compares to estimated enrollments of 1,827,000 in mathematics and statistics courses in these departments in fall 1985. Thus approximately 13% of the teaching in these departments was in computer science (the 1985 Annual AMS Survey results give an estimate of 10% rather than 13%, but the 1984 Annual AMS Survey yielded 12%—see Box 3.2).

In two-year colleges in 1985, approximately 10% of the 1 million students enrolled per term were enrolled in computing and data processing.

The 1985 CBMS Survey listed 27,500 bachelor's degrees awarded by departments of "mathematics" and 400 awarded by departments of statistics in the period July 1984 to June 1985. Of these 27,500 degrees, 40% were awarded in computer science (8,700) or jointly with computer science (2,500).

The 1985 CBMS Survey reported that of the 3,750 Ph.D.s on the nation's full-time computer science faculty, 41% had their doctorates in mathematics. Of the 2,200 Ph.D.s on the part-time computer science faculty, 61% had their doctorates in mathematics. Of the total full-time computer science faculty, 35% had their highest degrees in mathematics, and of the total part-time computer science faculty, 42% had their highest degree in mathematics. Of the 5,650 members of the full-time computer science faculty, 2,050 were employed by "mathematics" departments. Of the 5,350 members of the part-time computer science faculty, 3,350 were employed by "mathematics" departments. This translated to 3,150 full-time equivalents (FTE) of faculty members teaching one-half of the computer science sections in "mathematics" departments and 4,250 FTE of faculty members teaching the other half in computer science departments. It is noted that computer science departments are concentrated in the universities where teaching loads are lower.

of postdoctoral positions and no candidates from other disciplines who fit the faculty needs. Ad hoc hiring practices continued, partly because of a lack of suitable candidates for regular faculty positions.

From another perspective, by 1970 a large infrastructure of mathematical sciences graduate study and research had been established across the country and was spread through more than 150 universities. Success in research was clearly the principal criterion for respect within this community, and the research environment was clearly the best in the world. However, federal support for mathematical sciences research became less available, as did other governmental and institutional support (NRC, 1984), and the employment market was very depressed. There were many discouraged faculty members and persons seeking faculty positions. Many defected to other areas. By 1980, the mathematical sciences infrastructure was clearly weakening.

During the 1970s and continuing until the present, many departments' programs, especially at four-year colleges, contained a mixture of mathematics, statistics, and computer science. Planning was confounded further by conflicting trends within these three disciplines. Computer science was booming, statistics was growing steadily, and mathematics was struggling to adjust to a depressed employment market, fewer majors, and huge enrollments in introductory courses.

Computer science was emerging as a separate academic discipline. Many computer science programs had been formed within mathematical sciences departments, and the number of majors and the course enrollments were rising rapidly. Since there were far too few people with academic degrees in computer science to fill the available faculty positions and because of computer science's close connections to mathematics, many mathematics faculty members were able to cross over to computer science. And students who once might have been mathematics majors began to choose computer science as a major. By 1988 separate computer science departments had been established in most large universities, but in smaller institutions the hybrid department was still the rule. Approximately half of all computer science enrollments continue to be in these combined departments, thus competing for faculty time and energy and the interest of the students (see Box 1.1).

While much of the ferment over the past two decades also affected other academic disciplines, especially the sciences and engineering, the impact on the mathematical sciences was more extreme. The declines in the numbers of mathematics degrees awarded were relatively larger, and the declines in the numbers of majors in other science and engineering disciplines turned around much more quickly in the early 1980s. Mathematics has been the slowest field to recover—although there has been some recovery—one reason being the close ties between mathematics and academe. No other science or engineering discipline depends as heavily on academic employment for its graduates, especially those with doctorates, as does mathematics. Consequently the health of the mathematical sciences enterprise is very closely tied to the health of education, especially higher education.

By 1982 a number of serious problems posing a risk to the general health of mathematics had become apparent. The numbers of degrees awarded were near the lowest; course enrollments were the highest, with the heaviest concentration at the lower levels; teaching loads had increased dramatically; federal support for research was at a

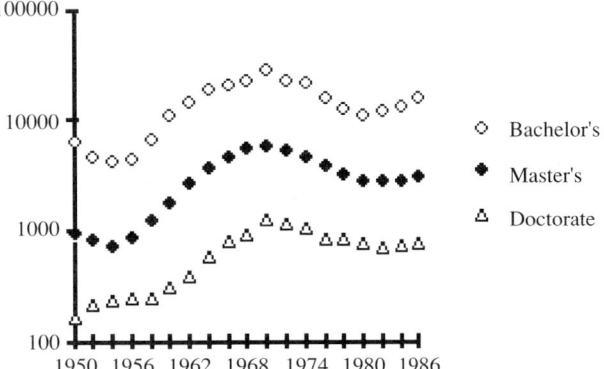

FIGURE 1.4 Mathematical sciences degrees awarded. SOURCE: National Center for Education Statistics (NCES, 1988a).

> **BOX 1.2 Sources of Data**
>
> Several sources of data were used to compile this report. The main sources include:
>
> - American Mathematical Society (AMS);
> - Conference Board of the Mathematical Sciences (CBMS);
> - National Center for Education Statistics (NCES) of the Department of Education;
> - National Research Council (NRC); and
> - National Science Foundation (NSF).
>
> In general, the mathematical professional societies' (AMS and CBMS) data relate only to the field of mathematical sciences and do not allow any comparisons across fields. When field comparisons are made, the sources of data are usually the NRC, NSF, or NCES. Inconsistencies do arise, partly because of different survey populations. For instance, some data on mathematical sciences include data on computer science. Where possible in this report, mathematical sciences data have been separated from computer science data. It is not feasible to reconcile or explain all the differences. Analysis of data in detail reveals differences that cannot be reconciled, but the implications of these differences appear to be minor. Nevertheless, the different sources have been found to be consistent enough to depict the general circumstances in the mathematical sciences enterprise.
>
> The tables in the text are numbered consecutively within each chapter, as are the figures (mostly graphs). Tables giving the data used to construct the figures presented in this report are included in the report's appendix and are numbered to correspond to the relevant text chapter rather than to a particular text figure or table. For example, Table A4.3 is the third table in Appendix Tables that contains data for Chapter 4, while Table 4.3 is simply the third table in the text of Chapter 4. The sources of the information shown in the tables and figures are given according to the standard referencing system used throughout the report.

very low point, especially in core areas of mathematics; and faculty morale was frequently low. Responding to the expansion of the previous two decades and the associated problems had taken most of the faculty time and energy. During that period, whole new areas of mathematical sciences had developed, including operations research, discrete mathematics, mathematical biology, statistical design and analysis, and nonlinear dynamical systems. In fact, the term "mathematical sciences" itself had become a part of the taxonomy of science. In spite of these new developments in the mathematical sciences, new applications, and new opportunities for using technological advancements, the teaching of mathematics had essentially not changed. Both the curricular content and its delivery had remained static. Mathematical sciences departments were not able to simultaneously cope with enormous instructional loads, maintain excellence in faculty scholarship, and allocate resources to innovations or even known improvements. The forces at work were too diverse and too disparate.

National Efforts Toward Renewal

In 1982 the mathematical sciences community began to address these problems on a national level. In 1984 the National Research Council (NRC) published *Renewing U.S. Mathematics: Critical Resource for the Future* (referred to as the David Report; NRC, 1984), documenting the weakening of federal support for research in the mathematical sciences. That report was the first of a series of efforts within the NRC and in professional societies to assess the health of the mathematical sciences and to design a plan for renewal. The NRC project Mathematical Sciences in the Year 2000 (MS 2000), of which this report is a part, was initiated in 1986. MS 2000 is an effort to assess the state of college and university mathematical sciences and to design a national agenda for revitalization and renewal. The events of the past three decades, detailed here, lend urgency to this effort. A major step in broadening the audience for this message and including all of mathematics education was taken by the NRC in publishing *Everybody Counts* early in 1989 (NRC, 1989). The issues and implications identified in this report and the two additional descriptive reports on curriculum and resources will assist the MS 2000 Committee in presenting an agenda that will ensure a healthy flow of mathematical talent into the next century.

Contents of This Report

Fundamentally this report concerns students and teachers. The events and forces described above indicate the complexity of this simple-sounding enterprise and how the current predicaments have developed. Box 1.2 describes the sources of the data used to compile this report and explains the relationship between the text tables and figures and the additional data presented in the report's Appendix Tables. Box 1.3 details characteristics of the statistics component of college and university mathematical sciences and describes the the context in which information in that area is provided. Chapter 2 describes in broad strokes the larger communities of the U.S. labor force and higher education, which both encompass the mathematical sciences enterprise. Chapter 3 describes the major components of, trends in, and utilization of college and university mathematical sciences. Chapter 4 focuses on mathematical sciences majors, both undergraduate and graduate. Chapter 5 describes mathematical scientists in the workplace; colleges and universities are a principal topic, since academe is still the dominant employer of mathematical scientists. That situation, however, is changing. The increased use in various professions of the mathematical sciences adds to their traditionally important uses in everyday life, civic activities, and our rich intellectual culture.

BOX 1.3 Statistics

The discipline of statistics is included in this report as a part of the mathematical sciences, principally because statistics has an intellectual base in mathematics, mathematics students are the principal source of statistics graduate students, and significant federal funding for academic research that develops fundamental statistical concepts and methods comes from the "mathematical sciences" units of federal agencies (NRC, 1984).

Degree programs in statistics are mostly graduate degree programs. The number of students enrolled in statistics and the number of undergraduate statistics majors are much smaller than the analogous numbers for mathematics. In major universities, statistics usually constitutes a separate academic department, but in other institutions statistics is likely to be taught in the same unit as mathematics (see Box 3.3). In addition, statistics courses are taught in a variety of administrative units, including business, engineering, medical sciences, and social sciences.

Partly because of the close administrative and intellectual ties between statistics and mathematics, much of the data on statistics in colleges and universities in this report is aggregated with analogous data on mathematics. Some disaggregation is possible and has been done when possible in this report. However, in general, the data are dominated by those for mathematics, and caution must be used in drawing conclusions about statistics from the aggregated data.

2 The U.S. Labor Force and Higher Education

- More new jobs will require more postsecondary mathematics education.

- The rate of growth in mathematically based occupations is about twice that for all occupations.

- Overall college enrollments are expected to decline until 1995.

- Minorities and women, now less likely to choose mathematically based occupations, will constitute larger shares of new workers.

- Shifting interests of college students and high attrition have reduced the number of students in the natural sciences and engineering pipeline.

Introduction

The need for a mathematically educated citizenry has grown steadily over the past century and has accelerated in the past two decades. This rapid growth is projected to continue into the next century. There are increased needs for mathematical knowledge and skills in all current areas of use—practical, civic, professional, and cultural—but the needs for future workers have received the most attention. Recurring predictions of the two chief requirements for future workers—higher levels of skills and adaptability—imply that more, and possibly different, postsecondary education and especially mathematics education will be needed.

These projections are noteworthy in view of the significant rise over the past two decades in the educational attainment of the civilian labor force. The average worker in the labor force 20 years ago had a high school diploma. Today the average worker's level of education has increased to include almost a year of college. The percentage of the labor force that has attended college nearly doubled from 22% in 1965 to 40% in 1984, and more than 50% of the new jobs created between 1985 and the year 2000 will require some college education (Figure 2.1). At the same time, the share of the current labor force with less than four years of high school has dropped from 43% to 20%.

Mathematical sciences education is needed by all individuals in the labor force, although to varying degrees. Problem solving and numerical reasoning are becoming essential in increasing numbers of jobs, and higher levels of mathematical competency are required of those involved in technological development and implementation. As developments within the mathematical sciences occur and their applications to many areas expand, more specialized mathematical knowledge becomes a prerequisite for mathematics-related professions.

Much of the added responsibility for meeting this larger need for mathematically educated people for the work force is borne by colleges and universities. The extent to which this need can be met is central to this report, and

A Challenge of Numbers

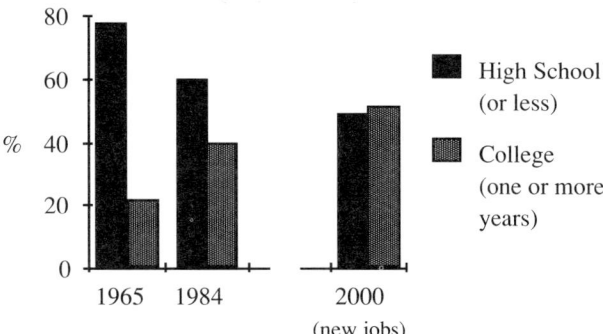

FIGURE 2.1 The educational requirements of the work force are increasing. SOURCES: Bureau of Labor Statistics (BLS, 1985) and Hudson Institute (BLS, 1987); see Appendix Table A2.1.

some of the problems to be overcome are new and formidable. By most assessments, the current flow of talent produced by U.S. college and university mathematical sciences programs is insufficient. Improving that flow will be complicated by a decreasing pool of U.S. students and by significant changes in the ethnic and racial composition of that pool, given a continuation of the current low interest in and attractiveness of the study of mathematics by groups that will have the most significant population growth. Many mathematical sciences programs are overextended and preoccupied with the large increases in remedial and precalculus enrollments of the past 15 years, and the teaching system is outdated, both in curricular content and in the methods and technology used in instruction.

More Skills and Greater Adaptability

The U.S. economy is projected to add 21 million new jobs between 1986 and 2000, after having added 31 million new jobs from 1972 to 1986, as reported in *Workforce 2000* (BLS, 1987). The report goes on to state: "For the first time in history, a majority of all new jobs will require postsecondary education" (BLS, 1987, p. xxvii). Adding the number of 1985 workers over age 50 who have more than four years of college work to the projected number of new jobs that will require more than four years of college yields more than 12 million jobs for college graduates by the year 2000. Assuming that current trends continue, this number is close to the total number of new graduates expected between 1985 and 2000. Thus the overall need for college graduates in the work force will be met if most of these college graduates enter the labor market and if their degrees are in the correct areas. However, these conditions are not likely to be satisfied, and significant shortfalls have been predicted in science and engineering. Recent National Science Foundation predictions point to a shortage of about half a million scientists and engineers by the year 2000, with shortages of 400,000 scientists and 275,000 engineers predicted for 2006 (AAAS, 1989). *Workforce 2000* (BLS, 1987) addressed the issue of requirements for increased skills by stating:

> The new jobs in service industries, all those available, will demand much higher skill levels than the jobs of today. Very few new jobs will be created for those who cannot read, follow directions, and use mathematics. . . . The fastest growing jobs will be in professional, technical, and sales fields requiring the highest education and skills level. Of the fastest-growing job categories, all but one, service occupations, require more than the median level of education for all jobs. Of those growing more slowly than average, not one requires more than the median education.
>
> Ranking jobs according to skills, rather than education, illustrates the rising requirements even more dramatically. When jobs are given numerical ratings according to the math, language, and reasoning skills they require, only twenty-seven percent of all new jobs fall into the lowest two skill categories, while 40 percent of current jobs require those limited skills. By contrast, 41 percent of new jobs are in the three highest skill groups, compared to only 24 percent of current jobs.

In meeting the projected needs for more highly educated workers, two problem areas have been noted. First, there is a growing mismatch between the emerging jobs, which will call for increasingly higher levels of skill, and the people available to fill them. Second, the labor market will be a place of churning dislocation, with companies coming and going and jobs changing and being redefined as the United States copes with rapid technological change and an increasingly competitive global economy. The ability of workers to adapt will be critical for success (BLS, 1988b; Richman, 1988).

Adaptability and education are virtually synonymous for workers. Acquired job-specific skills become secondary; knowledge, writing, problem solving, and numerical reasoning are critical. Better-educated workers already experience significantly shorter periods of unemployment after losing jobs. Unemployment rates of college-educated workers are approximately one-third the overall rate (NAS, 1987b).

Even within the same organization, adaptation to new work environments has become commonplace. To compete successfully, U.S. companies must be able to rely on workers to develop, learn, and adapt to new technologies. These abilities depend on the education of the workers, and since many new technologies are mathematically based, mathematics education is critical.

Growth in Science-Based Occupations

According to projections of the Bureau of Labor Statistics, eight of the ten fastest growing jobs will be in science-based occupations by 1995. Before 2000, industries will need many more computer programmers and operators, systems analysts, scientists, and engineers. The increase in the number of jobs requiring scientific or technical skills—many mathematics-based—is estimated to be significant and is predicted to occur at a rate much higher than that for all jobs (CPST, 1988). This projection is based on various circumstances pertinent to specific fields, including advances in technology and new applications, the increased importance of quantitative analysis in decision making, shortages of and replacements for doctoral degree holders, and replacements for people transferring to other occupations or retiring.

The projected increase in the demand for all scientists, engineers, and technicians between 1986 and 2000 is 36%, compared to a 19% increase in overall employment demands. For scientists the expected increase is 45% and for mathematical scientists 29%, compared to 76% for computer specialists, 32% for engineers, and 36% for technicians (NSB, 1987). These increases are projected from a 1985 base that was generally higher than the average base for the past 25 years.

The High Technology Recruitment Index, maintained by Deutsch, Shea & Evans, Inc., monitors demand for technical expertise based on the number of recruitment advertisements directed to scientists and engineers. Data have been collected since 1961, the year used as the base of 100. The index has averaged 106 and has ranged from a low of 44 in 1971 to a high of 158 in 1966. During 1987-1988, the index hovered at a moderately high level between 115 and 125, but it fell to about 100 in 1989.

In addition to there being a reasonably favorable outlook for mathematics-related jobs, such jobs are some of the most desirable, according to an article (Shogren, 1988) that reported on a study in the *The Jobs Rated Almanac*.

BOX 2.1 Degree Programs in Mathematics

Of the more than 3,300 higher education institutions in the United States, more than 2,500 have programs in mathematics. Approximately 1,000 of these are two-year institutions, and most of the remaining 1,500 offer programs leading to a bachelor's degree with a major in mathematics. About 425 institutions offer master's degrees in mathematics, and 155 offer programs for the doctoral degree.

Degree programs in mathematics education are frequently different from those in mathematics and may be located in a different administrative unit such as a college of education. Some institutions may offer separate degrees in applied mathematics or mathematical sciences, and joint bachelor's degrees in mathematics and computer science are becoming more common.

Several factors, only one of which is job outlook, are used to measure job desirability. Other factors are salary, stress, work environment, security, and physical demands. When these other factors are considered, the best 5 of 250 jobs rated—(1) actuary, (2) computer programmer, (3) computer systems analysts, (4) mathematician, and (5) statistician—are mathematics-based. Each of these requires an intensive mathematics background at the undergraduate level equivalent to a bachelor's degree in computer science, mathematics, statistics, or actuarial science (Shogren, 1988).

Much of the responsibility for meeting these challenges to provide workers with more and possibly different kinds of education rests with the U.S. higher education system.

BOX 2.2 Degree Programs in Statistics

The American Statistical Association (ASA) publishes lists of U.S. degree programs in statistics and in other areas with an emphasis in statistics (e.g., mathematics, business administration, and public health). The information below is taken from the 1987 list.

The degree programs are housed in 252 departments in 197 institutions as follows:

- 174 of the departments have names that include the designations statistics, mathematics, mathematical sciences, or combinations of these;
- 35 of the departments are biological units with degree programs titled biostatistics or biometry;
- 30 are departments of business administration; and
- 13 are scattered among agriculture, psychology, education, and engineering.

The list of programs includes:

- 131 bachelor's degree programs at 123 institutions;
- 217 master's degree programs at 172 institutions; and
- 164 doctoral degree programs at 122 institutions.

Higher Education in the United States

Higher education in the United States is extensive, diverse, and increasingly expensive. More than 64 million (about 1 of 4) people in the United States are involved in giving or receiving formal education at all levels. Of these, 14.4 million are involved in higher education at 3,340 institutions. These institutions spent an estimated $112 billion in 1986-1987. Such expenditures, and thus the costs of higher education, have increased significantly—in 1985, the cost to students of attending college was 3.5 times the cost in 1966 and twice the cost in 1975 (NCES, 1987a).

Of the 3,340 U.S. institutions of higher education, 1,309 offer less than four years of work, typically two years. Of the others, some offer as the highest degree the bachelor's degree (707), a first professional degree (93), a master's degree (566), some degree between a master's and a doctorate (153), and the doctorate (473). Some 37 institutions do not grant degrees (NCES, 1987a). Most of these institutions offer degrees in the mathematical sciences (see Boxes 2.1 and 2.2).

On average, of 100 people involved in higher education, 86 are students, 9 are administrative or support staff, and 5 are faculty members. In 1985 women students outnumbered men students by 6.4 million to 5.8 million (NCES, 1987a). Some 7.1 million were classified as full-time, while 5.1 million were classified as part-time. There were 10.6 million undergraduates and 1.3 million graduate students. Minorities made up 18% of the students in 1986 compared to 15% in 1976 (NCES, 1988a).

College enrollments peaked in 1983 at 12.5 million after increasing by 40% from 1970 to 1980 and increasing slightly in the early 1980s. During the period from 1970 to 1985, the percentage of adults with at least four years of college increased from 11% to 19%. Undergraduate enrollments have declined since 1983, but graduate enrollments have been steady since the 1970s, with small increases in the middle 1980s. Total undergraduate enrollment in colleges and universities increased 13.4% in the decade ending in 1986, while during this same period the total number of 18- to 24-year-olds decreased.

In 1984-1985, higher education institutions awarded

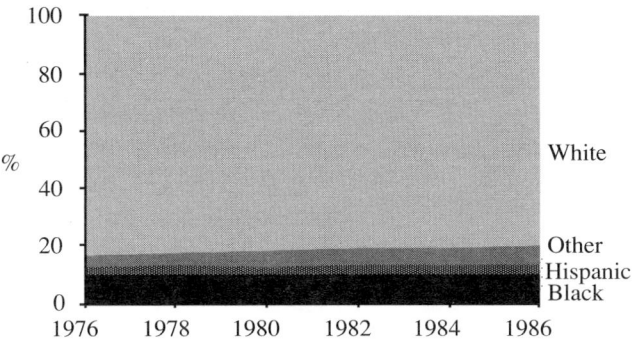

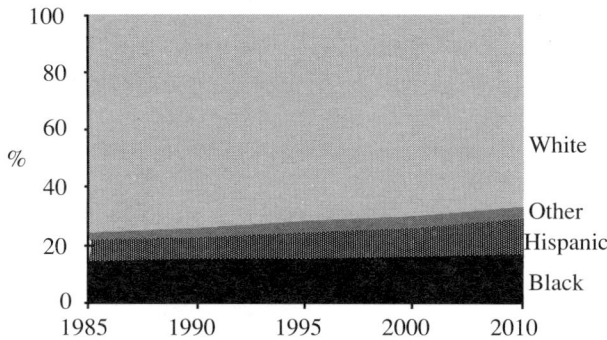

FIGURE 2.2 Percent distribution of undergraduate enrollments by race and ethnic group. SOURCE: National Center for Education Statistics (NCES, 1987a).

FIGURE 2.3 The pool of college students is changing, 18- to 24-year-old population. SOURCE: Bureau of the Census (BOC, 1986); see Appendix Table A2.2.

979,000 bachelor's degrees, 286,000 master's degrees, and 32,700 doctoral degrees. The most popular areas for the bachelor's degrees were business and management, engineering and engineering technology, social sciences, education, and the health professions (see Appendix Table A2.6). The leading areas for master's degrees were education and business and management, and for doctoral degrees were education, the social and behavioral sciences, the life sciences, the physical sciences, and engineering (NCES, 1987a).

The Pool of Potential Students and Workers

Between now and the year 2000, the U.S. population will grow more slowly than at any time since the 1930s, and the average age will increase to 36, six years older than the averge at any time in the history of the nation. More women will enter the work force, minorities will be a larger share of new workers, and immigrants will represent the largest share of the increase in the work force since World War I. In fact, native white men, constituting 47% of the 1985 labor force, will constitute only 15% of the new workers between 1985 and 2000 (BLS, 1987). If current trends continue, these projections indicate reduced numbers of both college students and persons choosing mathematically based occupations.

The traditional pool of college students, persons between the ages of 18 and 24, is shrinking, and the fraction of minorities in this shrinking pool is increasing. Minorities have been less likely than majority whites to enroll in college in the traditional age range of 18 to 24, but they are slightly more likely to enroll as older students.

College enrollments are expected to decline through the late 1980s and early 1990s for two reasons. First, the 18- to 24-year-old group in the U.S. population peaked at 30 million in 1981, is now declining, will reach a low of 24 million in 1995, and then will climb back to about the 1970 level by the year 2000 (BOC, 1986). The decline will not be uniform geographically but, in general, will take place north and east of a line extending from northern Florida to northern Idaho. South and west of that line, there will be increases. Geographic mobility will thus complicate the effects of the decline. Second, demographic and socioeconomic projections predict a population that will have a lower college attendance rate if current patterns persist (Figure 2.2). The fraction of the total 18- to 24-year-old population represented by blacks and Hispanics will increase from 22% in 1985 to 27% in 2000 and to 30% in 2010 (Figure 2.3), and these two groups have had a lower college-attending rate than has the general population.

The possible decline in enrollments is expected to be mitigated by two factors. Enrollments from the 25- to 34-year-old group are expected to stay strong. The fraction of this age group enrolled in education has approximately doubled from 1.6 million in 1970 to 3.2 million in 1985. This age group constituted one-quarter of all enrollments in 1985. The second factor is the growing enrollments of foreign nationals. The number of these students attending U.S. colleges and universities has increased by about 50% in the past ten years (NCES, 1987a).

In 1986 about one-third of the U.S. 18- to 24-year-olds who were high school graduates were enrolled in college, and this fraction represented about one of every four in the total cohort of that age. The relationship between this segment of the population and those enrolled in colleges and universities is not direct, for, in recent years, while the 18- to 24-year-old population was decreasing, enrollments in colleges and universities were increasing. However, the group aged 18 to 24 still enrolls in college at a rate more than three times the rate for the group aged 25 to 34 and remains the traditional and principal pool of college enrollees.

Of those who do not attend college, some will drop out of high school and others will graduate from high school but not enroll in college. In 1985 high school dropouts constituted 14% of 18- and 19-year-olds. This fraction was higher for Hispanics and blacks but lower for whites (Figure 2.4). Not only were blacks and Hispanics less likely than whites to graduate from high school, but also those who did graduate were less likely to enroll in college (Figure 2.5).

College participation rates for 25- to 34-year-olds who completed high school also showed a variation among racial and ethnic groups. In this age group 8-10% of high school graduates were enrolled in college. Blacks and Hispanics were slightly more likely than whites to be enrolled as older students.

Persistence in College Enrollment

Approximately 1 of 6 high school seniors persists through four years of college in the traditional pattern. This ratio is much lower for blacks (1 of 10) and for Hispanics (1 of 15). Data describing persistence indicate which students stay in the general education pipeline and, for those who do not stay in the pipeline, when they exit. Although persistence has been linked to degree attainment, the results should be interpreted with some caution. Persistence and traditional pattern here refer only to those who enter a four-year college following high school graduation and remain enrolled full-time. Several studies have shown that the majority of students who earn a bachelor's degree do deviate from this traditional pattern of enrollment. Delaying entrance, switching to part-time status, dropping

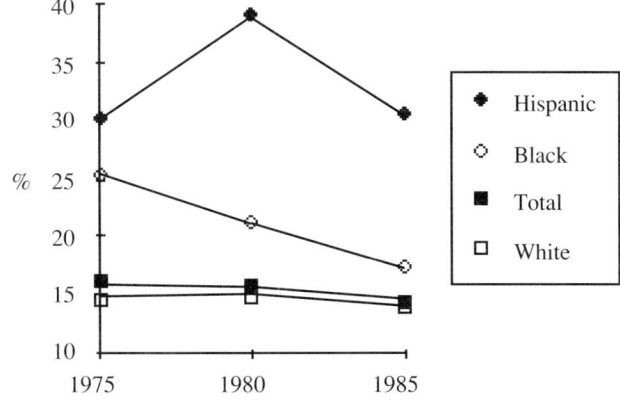

FIGURE 2.4 Percent of 18- and 19-year-olds who are high school dropouts, by ethnic group. SOURCE: National Center for Education Statistics (NCES, 1987a).

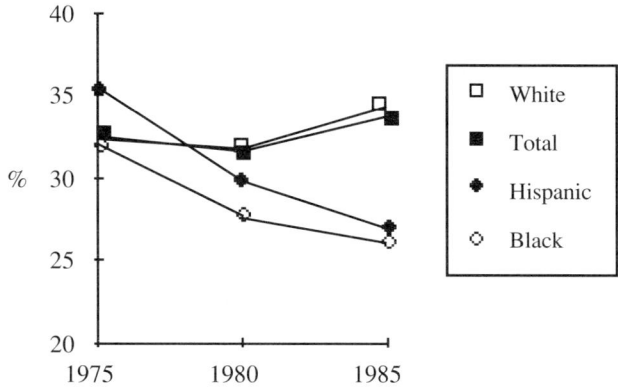

FIGURE 2.5 Enrollment in institutions of higher education as a percent of high school graduates. SOURCE: National Center for Education Statistics (NCES, 1987a).

out, taking a leave of absence, or transferring to a two-year college are some of the many ways students might diverge.

A group of 1980 high school graduates was surveyed by the U.S. Department of Education for six years following high school graduation (NCES, 1989). Of those surveyed, about one-third had never enrolled in postsecondary education. Blacks and Hispanics were less likely to have attended college; this tendency is also reflected in the different enrollment rates of 18- to 24-year-olds. Another one-third did start college, but not in the traditional way. Of the remaining one-third—those who did enroll full-time in college—about one-half persisted through four years of college, and three-fourths of these eventually attained degrees. Whites (56%) and Asians (61%) were more likely to persist after enrolling in college than were blacks (44%) and Hispanics (42%) (see Appendix Table A2.8). At no one point in the four years of college were students more likely to get off track. The lower persistence rates for blacks and Hispanics reflected the cumulative effect of fewer students continuing at each point (academic year and summers) rather than high attrition at any one identifiable stage (NCES, 1989).

Shifting Interests of College Students

In the last 20 years student interest has shifted dramatically among various college academic majors. This shift has been monitored in three different ways: (1) surveys of entering freshman on probable major and career, (2) enrollments in courses by field, and (3) degree production by field. Each of these measures points to similar trends. Students are more interested in fields of study that are job related, and of these job-related fields, the higher paying ones are more popular.

The proportion of entering freshman intending to major in business, computer science, and engineering has increased in the last 20 years, while at the same time freshman interest in the sciences, the humanities, education, and mathematics has decreased, in some cases dramatically. The declining interest in education is an exception to the general trend of the growing popularity of professional fields; factors such as low salaries, poor working conditions, low esteem, and broader career options for women have more than offset any increased interest in education. According to the Cooperative Institutional Research Program (CIRP), which conducts annual surveys of entering college freshmen, career-related fields have gained in popularity at the expense of virtually every field traditionally associated with a liberal arts education. A comparison of anticipated majors of college freshman with the distribution of baccalaureate degrees conferred confirms this shifting of interests (Figure 2.6). There is a strong correlation between the anticipated major and the general distribution of degrees conferred, even though there is considerable shifting of majors after students enter college.

The increased interest in marketable skills has raised serious issues and basic questions about the purpose of higher education that are beyond the scope of this report. However, these issues do concern the mathematical sciences because one aspect of the lack of interest in the mathematical sciences is the more general lack of interest in a traditional liberal arts education. The cutting issue raised by Astin et al. in *The American Freshman: Twenty Year Trends* (CIRP, 1987b, pp. 26-27) and a host of others concerned with higher education and its future is "whether the higher educational community should adapt passively to these 'market' trends in student expectations, or whether the inherent dangers in such trends should be recognized and curricula revised accordingly. Should colleges simply phase out their programs in the humanities, cut back on their social science and education programs, and expand their offerings in business and technology?"

Data on enrollments in courses by major field of study display the same general trends of shifting interest as do data on the intended majors of freshmen, in spite of the numerous changes in major made during college. Enrollments and degree production by field generally have mirrored what students say they are interested in as entering freshman. In 1982 the most popular areas in terms of enrollments by field were business and commerce and combined engineering and computer science. These three fields accounted for one-third of all enrollments. The number and the distribution of baccalaureate degrees conferred reflect the general enrollment patterns.

From 1971 to 1985, the total number of bachelor's degrees conferred increased 17%, from about 840,000 to 979,000. Business and management, computer and information sciences, engineering, and the health sciences all showed remarkable and consistent increases in the number of degrees conferred and also accounted for a larger share of all degrees conferred. The number of education (-50%), English (-47%), mathematics (-39%), and social sciences (-32%) degrees each declined precipitously. And fine arts, the life sciences, physical sciences, and agriculture showed either a mixed or a fairly constant production of bachelor's degrees (Figure 2.6).

With the growing popularity of certain majors and the loss of appeal of others, the distribution of bachelor's degrees conferred has changed dramatically since the early 1970s. In 1985 the distribution of degrees conferred was roughly 25% in business and management and 10% each in the social sciences, engineering, education, and the humanities, including English; the physical sciences, mathematics, computer science, and the life sciences each accounted for less than 5 percent of the degrees conferred.

Natural Sciences and Engineering

The total production of natural sciences and engineering bachelor's degrees has grown steadily in the last 25 years, but there have been wide differences in the changes in the various fields. The category natural sciences and engineering includes the fields of physical, mathematical, life, and computer sciences and engineering but does not include the social sciences. Since the early 1970s, the number of degrees awarded has grown considerably in engineering and computer science, has remained relatively constant in the physical sciences, and has plunged in mathematics. Aggregate data on science and engineering degrees masks these important differences between the subfields. A comparison of the total number of people in the natural sciences and engineering pipeline with degree production at different levels in various fields highlights some of these differences.

In 1985 natural sciences and engineering degrees accounted for 212,300 of the 979,000 bachelor's degrees conferred, or about 22%. Trends in the production of natural sciences and engineering degrees have followed the same general pattern described above—job-related degrees have increased in the last 10 to 15 years, and those in the arts and sciences and not specifically job related have either remained constant or have decreased.

For students continuing on to doctoral degrees, conferral of a bachelor's degree can be viewed as a midpoint in the educational process. From this viewpoint, the pipeline begins seven to eight years earlier in high school. At this critical stage students either take the requisite courses to continue in the pipeline or they drop out of the science and engineering track. At various points in the pipeline, losses occur, and students are not likely to return once they have left. After the conferral of the bachelor's degree, seven to eight years are required to complete work for the doctoral degree for those who do continue.

The time required to educate a scientist or an engineer can extend to at least 15 years from the time a student first has some choice in the selection of courses in high school to the awarding of the doctoral degree. As an illustration of this lengthy and leaky process, consider the fact that of 1,000 students who were high school sophomores in 1977, only 2 will have continued in the pipeline to receive a doctoral degree in science or engineering by 1992. Of these 1,000 high school sophomores, 180 were interested in science or engineering as sophomores, but by the time they were seniors only 150 were still interested. Only slightly more than half of these, 85, continued on to college with the intention of majoring in a science or engineering field. Approximately 51 of these 85 received a bachelor's degree in a natural science or engineering field. Of those who received bachelor's degrees, about 15 continued on for graduate work, and 11 of the 15 received a master's degree. Of these 11 who remained in the system, 2 will continue their studies to successfully complete a doctoral degree (NSF, 1987b).

Analysis of degree production for select science and engineering fields shows different patterns for the various fields. The time elapsed from receiving the baccalaureate to earning the doctorate has recently lengthened slightly and shows a variance by field, ranging from a low of 5.6

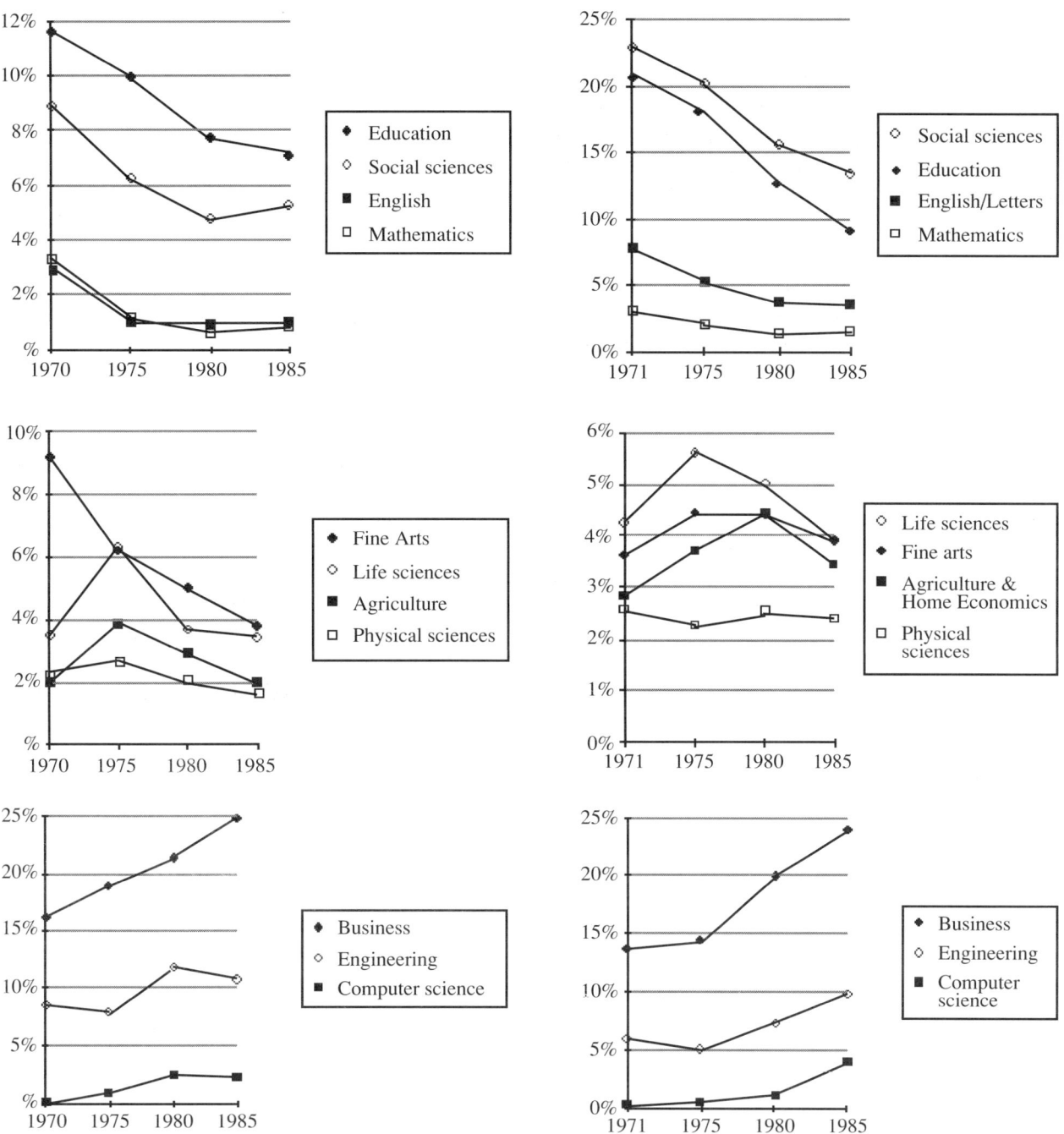

FIGURE 2.6 Shifting interest in selected majors. Left: Anticipated college major of entering freshmen. Right: Bachelor's degrees of exiting seniors. SOURCES: Cooperative Institutional Research Program (CIRP, 1987a) and National Center for Education Statistics (NCES, 1987b).

years in 1970 for chemistry to a high of 11.9 years in 1986 for the health sciences. For most fields, the time to complete a doctorate after receipt of a baccalaureate has been between 6 and 8.5 years. For purposes of simplifying analysis between fields, 7 years from receipt of the baccalaureate to receipt of the doctorate, 5 years from the master's to the doctorate, and 2 years from the baccalaureate to the master's are used below as averages for all fields.

The numbers of degrees conferred from 1971 to 1985 in selected fields for the three different levels—doctorate, master's, and baccalaureate—were analyzed to compare attainment percentages from one degree level to a higher level. Allowing for the lags of 2, 5, and 7 years, the total number of degrees at a higher level was divided by the total number at a lower level to give an attainment percentage. Those percentages are given in Table 2.1 (see Appendix Tables A2.10 and A4.11 for more details). There is no adjustment for entry into degree programs by students from outside the United States, as there is none for U.S students changing fields between degrees. Since about half of the doctorates in engineering and in the mathematical sciences are awarded to non-U.S. students, adjustments recognizing this would significantly lower the analogous attainment rates. Taken for one field alone, these rates are not very meaningful, but comparisons between fields are of interest. In the mathematical sciences, since the numbers of degrees awarded at all three levels have increased and decreased together—that is, the lags have not been meaningful—the rates are less significant. However, comparisons between fields reveal more similarity between the mathematical sciences and engineering than between the mathematical sciences and the other sciences. And attainment rates for advanced degrees are lower for the mathematical sciences than are those for all of the natural sciences and engineering.

The Challenges and the Responsibility

The needs of the nation's labor force, the shrinking and changing pool of workers, the shifting interests of students, and the projected shortages of scientists and engineers provide a matrix of challenging numbers for U.S. higher education. The major responsibility for meeting the challenges rests with the mathematical sciences component of higher education, the subject of Chapters 3 and 4.

Mathematics has always been a major part of higher education, but its fundamental role in society has expanded significantly in recent years. The complex circumstances in higher education and in the work force described above have combined with equally complex circumstances within the mathematical sciences to produce layers of formidable and interconnected problems that must be solved to meet the nation's needs for mathematically educated workers.

TABLE 2.1 Attainment rates of advanced degrees for selected fields, 1971 to 1985

	Master's/Bachelor's (2-year lag)	Doctoral/Master's (5-year lag)	Doctoral/Bachelor's (7-year lag)
All natural science and engineering	22%	21%	5%
Engineering	33%	17%	6%
Life sciences	14%	54%	8%
Physical sciences	25%	56%	15%
Mathematical sciences	21%	18%	4%

SOURCES: National Center for Education Statistics (NCES, 1987) and National Science Foundation (NSF, 1987b).

3 College and University Mathematical Sciences

- College and university mathematical sciences constitute a vast and diverse system that accounts for approximately 10% of all higher education.

- The system is strong at the top but is weakening at all levels.

- Precollege indicators predict mild improvements after a long decline.

- The transition from high school to college mathematics is one of the most troublesome in education.

- Enrollments in mathematical sciences courses have doubled in the last 20 years, but the increases have all been at the lower levels, with remedial enrollments leading the way.

Introduction

The academic mathematical sciences consist principally of mathematics and statistics. Also included are programs labeled applied mathematics, many areas of applied statistics, and the more mathematical parts of operations research, mathematical biology, engineering, and economics. The boundaries are by no means distinct. For example, it would be very difficult to determine any reasonably precise boundary between applied mathematics and theoretical physics or between mathematics and computer science. As recently as ten years ago, computer science was frequently included in the mathematical sciences, but that is no longer the case. Academically, the boundaries are not important and in fact are better disregarded. However, for description, administration, and policy, general boundaries need to be understood.

The diversity of the profession is illustrated by the large number of professional organizations that have an interest in college and university mathematical sciences (see Box 3.1), and much of the information about the people in the mathematical sciences comes from the professional organizations, in particular, the annual surveys conducted by the American Mathematical Society (AMS) and the surveys of the Conference Board of the Mathematical Sciences (CBMS). Boxes 3.2 and 3.3 describe the nature of these surveys.

College and university mathematical sciences in the United States constitute a vast enterprise with serious and diverse responsibilities that are critical to the welfare of the nation and to the maintenance of at least the disciplines of mathematics and statistics. There are mathematical science programs in at least 2,500 institutions of higher education, and these provide nearly 10% of all the teaching in U.S. higher education and approximately 30% of the teaching in the natural sciences and engineering. Each term, approximately 3 million students are taught by approximately 50,000 teachers, about 27,500 of whom are full-time, 14,500 are part-time, and 8,000 are graduate assistants.

Of the 2,500 institutions that have mathematical sciences programs, about 1,000 are two-year institutions, another 1,000 offer a bachelor's degree as their highest mathematical sciences degree, nearly 300 offer master's degrees as their highest degree, and nearly 200 offer a doctoral degree in the mathematical sciences. (Most of the institutions with master's and doctorate programs also have a baccalaureate program.) Many of these graduate institutions have departments and degree programs in each of mathematics and statistics. Some also have separate programs in applied mathematics or operations research. In addition to these programs, many institutions have programs closely related to the mathematical sciences in other academic units, for example, biostatistics or biometry in health science areas; operations research in engineering; business statistics, management science, or econometrics in business; statistics in social sciences; and mathematics education in education (see Boxes 2.1 and 2.2 for a breakdown of institutions by degree program for mathematics and statistics).

Most of the institutions granting bachelor's degrees, some granting master's degrees, and a few granting doctoral degrees have only one department in the mathematical sciences, and that department frequently houses a program in computer science. In the two-year institutions, the unit that contains the mathematical sciences may contain other areas of science or technology.

The responsibilities of college and university mathematical sciences are broader than those of any other academic area. These diverse responsibilities include providing courses for general education, service courses for other disciplines, programs for middle and secondary school mathematics teachers, and courses for elementary school teachers; educating college and university mathematical sciences faculty members, mathematical science researchers, and applied mathematical scientists; and nurturing the continued development of the disciplines of mathematics and statistics.

Strong at the Top

Mathematical sciences education and research at the highest levels in the United States are generally considered to be the strongest in the world. This strength comes from both the education of U.S. students as researchers and the immigration of mathematical scientists into the United States. The U.S. environment for mathematical research is clearly one of the best in the world. But a major concern of the mathematical sciences community, a concern that has far-reaching consequences, is how to preserve this strength. Edward E. David, Jr., has summarized the current situation as follows: "American mathematics is strong—way out of proportion to its numbers, way out of proportion to its level of support today. But will it be able to sustain and renew itself in the future? Unfortunately that problem has not gone away. It is in fact more pressing than ever" (AAAS, 1988b).

Winners of the Fields Medal, the world's most prestigious award for research in mathematics, have often been mathematicians from the United States. Awarded to outstanding research mathematicians every four years since it was established in 1936 by Professor J. C. Fields to recognize existing work and the promise of future achievement, this monetary prize and medal usually go to mathematicians who are less than 40 years old. Of the 30 winners of the Fields Medal, 9 were born in the United States and an additional 7 were affiliated with U.S. institutions at the time they won the award.

Despite such positive indicators of the position of U.S. mathematical sciences research in the world, several other indicators have implications that are mixed or inconclusive. Prestigious awards, increased collaboration, and development of new fields are signs of the vitality and strength of this enterprise, but the share of publications and citations attributed to U.S. mathematicians has been steadily dropping, in fact, dropping faster than that for any recorded U.S. research field.

Scholarly productivity is probably more narrowly and rigidly defined in mathematics than in any other science and engineering field. The productivity of research scientists has been assessed by monitoring (1) the number of articles published and (2) the number of times these articles are cited. In 1984, U.S. researchers produced 37% of the world's research articles in mathematics. This compares to

BOX 3.1 Professional Organizations

The seven general professional organizations whose primary interests are college and university mathematical sciences are the following:

• American Mathematical Association of Two-Year Colleges (AMATYC). Established in 1975, AMATYC's interests are, as the name implies, the mathematics and professional issues in two-year colleges. AMATYC currently has approximately 1,900 members.

• American Mathematical Society (AMS). Established in 1888, AMS's interests have centered on research and graduate study in mathematics. AMS currently has approximately 23,000 members.

• American Statistical Association (ASA). Established in 1839, ASA's interest is general statistics, including mathematical statistics and applications in various disciplines. ASA currently has approximately 15,000 members.

• Institute of Mathematical Statistics (IMS). Established in 1930, IMS's interest is mathematical statistics. IMS currently has approximately 3,000 members.

• Mathematical Association of America (MAA). Established in 1915, MAA's interests have centered on issues in undergraduate mathematics. MAA currently has approximately 27,000 members.

• National Council of Teachers of Mathematics (NCTM). Established in 1920, NCTM interests have centered on the teaching of mathematics, both at the college and precollege levels. NCTM has approximately 76,000 members, of whom about 38,000 are secondary school teachers and 3,000 are college faculty members.

• Society for Industrial and Applied Mathematics (SIAM). Established in 1952, SIAM's interests have centered on research and applications of mathematics. SIAM currently has approximately 7,000 members.

There is overlap in the memberships of these professional societies. The four college and university mathematics societies, AMATYC, AMS, MAA, and SIAM, have a combined membership of approximately 46,000 people, and the two statistical societies, ASA and IMS, have a combined membership of approximately 17,000 people.

In addition to these seven, there are other organizations that have specialized interests in college and university mathematical sciences. These include the Association for Symbolic Logic (ASL), the Association for Women in Mathematics (AWM), the Biometric Society, the Econometric Society, the Fibonacci Association, the National Association of Mathematicians (NAM, which is concerned with the interests of blacks in mathematics), the Operations Research Society of America (ORSA), the honorary society Pi Mu Epsilon, and The Institute of Management Sciences (TIMS).

Four confederations of professional organizations have some of the above organizations as members. The Joint Policy Board for Mathematics (JPBM) represents the AMS, MAA, and SIAM. The Conference Board of the Mathematical Sciences (CBMS) represents the following 15 professional societies: AMATYC, AMS, ASA, ASL, AWM, IMS, MAA, NAM, NCTM, SIAM, the Association of State Supervisors of Mathematics, the National Council of Supervisors of Mathematics, ORSA, the Society of Actuaries, and TIMS. The Council of Scientific Society Presidents includes representatives from 33 organizations, including AMATYC, AMS, CBMS, MAA, NCTM, and SIAM. The Commission on Professionals in Science and Technology includes the AMS, MAA, and SIAM in its membership of 16 professional societies.

the 35% of all the world's scientific and technical articles produced by U.S. scientists and engineers. However, the U.S. share of mathematics articles has dropped significantly since 1973, when the share was 48%. The current share, 37%, is below the analogous fractions for clinical medicine (41%), earth and space sciences (41%), engineering and technology development (40%), and biomedicine (39%) and above those for biology (37%), physics (27%), and chemistry (21%). The share for all these fields has either dropped or is the same as in 1973, but the drop for mathematics has been the largest.

This drop in the U.S. share of mathematics articles is probably reflected in the growing percentage of references in U.S. articles to articles from other countries. That percentage increased from 16% in 1974 to 29% in 1984. All other fields had analogous increases over this period, but, again, the increase for mathematics was the largest.

Also, the influence of U.S. articles as measured by citations in the world's literature dropped slightly between 1973 and 1982. Even though the drop was slight, a much larger drop of 25% occurred in the rate at which U.S. articles were cited in non-U.S. articles. The citation rate in the world's literature was bolstered by a 23% increase in the rate at which U.S. articles were cited in U.S. articles.

There is evidence that international collaboration and university-industry collaboration are increasing in mathematics research. University-industry coauthored papers, as a percent of all industry mathematics papers, increased from 28% in 1973 to 42% in 1984. Internationally coauthored papers, as a percent of all mathematics papers with authors from more than one institution, increased from 34% in 1973 to 48% in 1984 (NSB, 1987).

Mixed Precollege Indicators

While the achievements of U.S. research mathematicians compare well internationally, as does the preparation of top U.S. students, the achievements of most U.S. students at the high school level do not. Recent international comparisons of achievement test scores in precollege mathematics place U.S. students well below those in countries that are now major economic competitors of the United States. For example, in a 1981-1982 test of students from 13 countries, the most able U.S. students taking the test (the top 1 percent) scored the lowest in algebra among the analogous cohorts of all 13 countries and among the

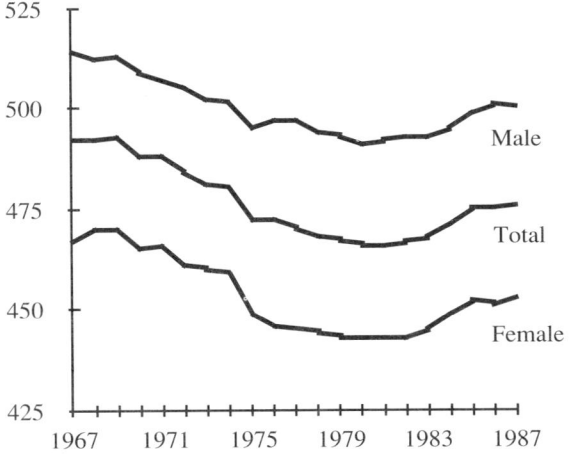

FIGURE 3.1 SAT mathematics scores, 1967 to 1987. SOURCE: College Entrance Examination Board as reported in Digest of Education Statistics (NCES, 1987b).

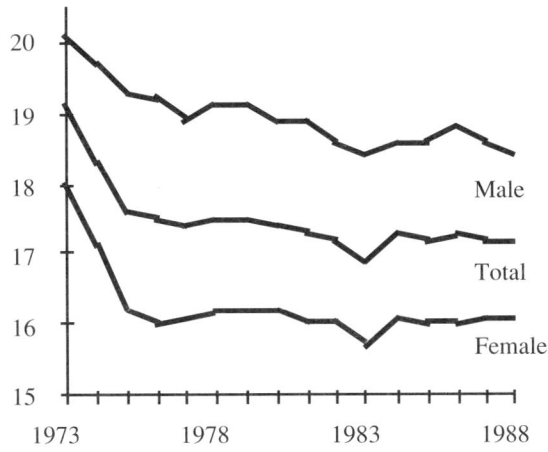

FIGURE 3.2 ACT mathematics scores, 1973 to 1988. SOURCE: American College Testing Program (ACT, 1989).

lowest in calculus. The algebra achievement of the U.S. top 5% was lower than that of the corresponding competitors from all but one country. The most able Japanese students scored higher than their counterparts in the other countries, and the average Japanese student outperformed the top 5% of the U.S. students in college preparatory mathematics (IAEEA, 1987).

Another more recent international assessment of mathematics and science skills places U.S. students last in mathematics performance overall among 13-year-old students in five countries and four Canadian provinces. On each of the six topics measured, U.S. students scored 10 to 20 percentage points below the top scorers. Yet when asked if they were good at mathematics, two-thirds of the American students felt they were, compared with one-fourth of the South Korean students, who were the top scorers (ETS, 1989).

Attitudes in the United States toward mathematics are mixed, and there are continuing myths about both the nature of mathematics and how one learns mathematics. Many believe that mathematics is a static subject and that success depends more on talent than on effort. Attitudes reported by U.S. students in grades 8 and 12 (see Appendix Table A3.1) reveal that almost two-thirds of U.S. students at both grade levels think mathematics and its subtopics are important, but only half of these students have an ease with mathematics, and less than half like mathematics. These attitudes toward mathematics at the high school level translate to fewer college freshmen who are interested in and prepared to pursue studies in mathematics.

Students are not learning enough mathematics in high school to prepare themselves for college-level courses or for the future workplace. White males are still leaving high school better prepared than either females or minority group members; however, this gap has been closing in recent years. The average number of 1-year course credits in mathematics completed by the high school seniors of 1982 was 2.5. For males the average was 2.6; for females, 2.5; for whites, 2.6; for blacks, 2.4; for Hispanics, 2.2; for Asians or Pacific Islanders, 3.1; and for Native Americans, 2.0. Half the students earned less than two credits in college preparatory courses (which include algebra 1, 2, and 3; geometry; trigonometry; analytic geometry; linear algebra; probability and statistics; and calculus). Nearly 1 in 20 earned less than one mathematics credit (NCES, 1985). Those students who plan to get a bachelor's degree do take more mathematics credits (3.1) than does the average high school graduate, and almost 90% of college freshmen have taken at least three years of mathematics.

Mathematics preparation in high school is a major factor in determining how well students perform on college achievement tests. After a general decline in the 1970s and early 1980s, average mathematics scores for the Scholastic Aptitude Test (SAT) have risen slightly and for the American College Testing (ACT) have leveled off in recent years (see Figures 3.1 and 3.2). Scores showing both gender and ethnic and racial differences have fueled the controversy over whether the tests are biased against women and minorities. Males have consistently scored higher than females by about 50 points on the mathematics section of the SAT for the last two decades; mathematics scores for blacks and Hispanics showed steady improvement during this same period but were below the national average (see Appendix Table A3.2).

The reversal of the steady slide in mathematics scores is cause for some optimism, but the improvements are not substantial enough. Students are still not well prepared for higher-level mathematics courses, and, according to the National Assessment of Educational Progress (NAEP), the progress that has been made is in lower-level skills. The NAEP has measured achievement in mathematics by U.S. students of ages 9, 13, and 17 in 1978, 1982, and 1986 and has extrapolated the assessment back to 1973 from previous NAEP analyses (see Appendix Table A3.4). The highlighted summary from the 1986 report includes the following (ETS, 1988):

Recent national trends in mathematics performance are somewhat encouraging, particularly for students at ages 9 and 17. Subpopulations of students who performed comparatively poorly in past assessments have shown significant improvement in average proficiency since 1978: at all three ages [9, 13, and 17], black and Hispanic students made appreciable gains, as did students living in the Southeast.

While average performance has improved since 1978, the gains have been confined primarily to lower-order skills. The highest level of performance attained by any substantial proportion of students in 1986 reflects only moderately complex skills and understandings. Most students, even at age 17, do not possess the breadth and depth of mathematics proficiency needed for advanced study in secondary school mathematics.

The lack of improvement in precollege mathematics preparation and the increased enrollments in college mathematics courses have led to difficulty for students in meeting the expectations of traditional college courses. This has resulted in fundamental and extensive changes in college and university mathematics.

Troublesome Transitions from High School to College

In general, there appears to be a mismatch between the articulated expectations of colleges and universities and the preparation of entering freshmen. Colleges and universities have not clearly articulated and enforced standards and have tried to accommodate extremely diverse backgrounds. A high school graduate, regardless of courses taken, can usually find a place in some college or university. This clash between expectation and preparation is evident in public and private statements and in the growing overlap between material covered in college courses and that covered in high school courses. In no discipline is this more apparent than it is in mathematics. Several authors and studies have addressed this issue, and problem resolution is generally considered a shared responsibility as reflected in the following selected positions (CARN, 1983, pp. 7-9):

> Right now, the colleges are genuine in their feelings that too many students are not adequately prepared for higher education. On the other hand, if the colleges had a modicum of conscience they must know that their own shift in standards and requirements had something to do with the situation that faces the schools today.

The fact is that across the board, not just at community colleges, college entrance requirements place little, and in some cases no, emphasis in the substantive content of what high school students should have mastered as the necessary prerequisite to college study. There is no common body of knowledge, no specific set of intellectual skills against which students can measure their own readiness or on which colleges can base admission and placement decisions.

Surprisingly, the courses a student takes are not important in getting into most colleges although they may be critical to success once a student is there. Half the colleges set no specific course requirements at all and only about one-fourth consider courses the students took in making the decision for or against admission. When specific courses are identified, the most frequently required courses are English (the usual requirement being four years), mathematics (two years is the average requirement) and the physical sciences (one year).

The transition from high school to college mathematics is especially troublesome for many students and institutions. Students enter college with widely varying levels of mathematical preparation: some are not competent in computation at a sixth grade level, and others have taken calculus in high school. Initial placement in college mathematics is complicated by this imbalance in the preparation of students and the several possible entry points for beginning students. Students begin mathematics in colleges with courses ranging from arithmetic to courses that assume mastery of calculus. It is not difficult to describe a dozen or so possible first courses within this range, and many institutions have as many as a half-dozen entry courses.

Placement programs have become more common and sophisticated over the past decade. In the mid-1970s the Mathematical Association of America (MAA) began its

BOX 3.2 AMS-MAA Survey Reports

The American Mathematical Society (AMS) has sponsored and conducted surveys of college and university mathematics programs and departments each year since 1957, and was joined in sponsorship by the Mathematical Association of America (MAA) in 1987. These surveys, initially covering only salaries, have expanded to include course enrollments; numbers and characteristics of faculty members; numbers, employment, and characteristics of new doctoral degree holders; faculty mobility; and nonacademic employment. The AMS-MAA survey reports are published annually in the Notices of the American Mathematical Society (AMS, 1976 to 1988).

The survey population of the AMS-MAA surveys is partitioned by degrees awarded in the mathematical sciences into groups as follows:

- Groups I, II, and III: These are departments that offer doctoral degrees in mathematics and that have been placed into one of three categories by their ranks in a 1982 assessment of research-doctorate programs in mathematics by the Conference Board of Associated Research Councils. Group I consists of the 39 top-ranked programs (those with an assessment rating between 3.0 and 5.0); Group II, the next 43 (those with a rating from 2.0 to 2.9); and Group III, the remaining 73 programs. It should be noted that many of the departments that offer these doctoral programs in mathematics contain programs in other areas of the mathematical sciences, most notably in statistics; and almost all have bachelor's and master's degree programs in mathematics. The AMS-MAA survey results cover all mathematical sciences programs in these departments, not just the doctoral programs in mathematics.
- Group IV: This group consists of 69 departments (or programs) of statistics, biostatistics, or biometrics that offer a doctoral program.
- Group V: This group consists of 57 departments (or programs) in applied mathematics, applied science, operations research, or management science that offer a doctoral program.
- Group VI: This group consists of 28 Canadian departments (or programs) that offer a doctoral program.
- Group M: This group consists of 273 departments in the mathematical sciences granting a master's degree as the highest degree. Most of these offer bachelor's degree programs, too.
- Group B: This group consists of 950 departments in the mathematical sciences granting a bachelor's degree as the highest degree.

It is noted that some parts of the AMS-MAA surveys have included Canadian institutions, and some of the AMS-MAA survey results include counts of Canadian degrees, which usually amount to 6-8% of the total of U.S. and Canadian degrees. Because the intent of this report is to describe the circumstances in U.S. institutions, the Canadian data are not included, when feasible.

Although programs in two-year colleges have not been included in recent AMS-MAA surveys, they were included in surveys conducted from 1977 to 1980.

Placement Test Program, which produces packages of tests for use in placing students in the initial college mathematics course. Scores on placement tests, SAT or ACT scores, the high school record, the college major, and student attitudes are variables that are used to determine initial placement. Some institutions have compulsory placement, but most are advisory. The following summary of the preparation of college freshmen in New Jersey demonstrates the magnitude of the placement problem and the challenge it presents to institutions (CARN, 1983, p. 12):

> In New Jersey, all freshmen entering public colleges and universities are tested in basic skills. Of the approximately 30,000 students who took the tests in

---

BOX 3.3 CBMS Surveys

The Conference Board of the Mathematical Sciences (CBMS) has sponsored five surveys of undergraduate programs in the mathematical and computer sciences, one every five years beginning in 1965. The surveys have sampled programs in universities, four-year colleges, and two-year colleges to project total enrollments in various courses and the numbers, responsibilities, and characteristics of faculty members. In the 1985 survey, computer science programs were treated separately, and attempts were made to separate the data on computer science programs that are located in mathematical sciences departments. Recent surveys have also asked for opinions on issues judged to be important by departments.

The 1985-1986 CBMS Survey population is based on the 1982 NCES classification of 157 universities (95 public and 62 private), 427 public four-year colleges, 839 private four-year colleges, and 1,040 two-year colleges. In this system of classification, universities are institutions that place considerable emphasis on graduate instruction. There were 156 institutions so classified in 1986. This group of institutions has a large overlap with the 155 institutions that offer doctoral degrees in mathematics. This latter group is used as a subpopulation in the AMS-MAA surveys, and for most purposes, assuming that these groups are the same causes no difficulty.

Many departments in the mathematical sciences contain programs in various areas: mathematics, applied mathematics, statistics, computer science, operations research, and others. Separate data on computer science were not commonly available until about 1980. In this report an attempt is made to separate the descriptive data into at least that for mathematics and that for statistics and to omit the data on computer science except as it affects mathematics and statistics.

The 1985-1986 CBMS Survey used the term "mathematics department" when referring to the unit in the mathematical sciences in an institution that might or might not have separate statistics or computer science departments. This "mathematics department" might have programs in all three areas—mathematics, statistics, and computer science—and in other areas. The CBMS survey population indicates that among the 157 universities, 40 have separate statistics departments and 105 have separate computer science departments. Among the 1,265 four-year colleges, 291 have separately surveyed computer science departments and only 5 have separately surveyed statistics departments. Very few of the 1,040 two-year colleges have separate units in either computer science or statistics. This indicates that most departmental units in the mathematical sciences teach the three disciplines of mathematics, statistics, and computer science. Of course, as the ASA list of statistics programs indicates, programs in statistics occur in several different units in colleges and universities, and many of these are not in the CBMS survey population. Similar circumstances exist in academic programs in computer science.

BOX 3.4 Minorities and Women

Fewer blacks, Hispanics, and women study mathematics and choose mathematically based careers than one would expect from their fraction of the total population. Asian-Americans choose mathematically based careers at rates higher than one would expect from their fraction of the total population. The number of Native Americans choosing such careers is very small, but the rate nearly equals the fraction this group forms of the total U.S. population.

The low numbers of blacks and Hispanics hold throughout the system—at all levels of the educational pipeline and in the workplace. The loss of women is more acute at the higher levels of the educational pipeline and in the workplace. These circumstances are well documented in the literature and by the data given in this report (see Chapter 4), but the low numbers are not described repeatedly for each group or educational activity. Instead, unusual patterns and significant situations pertaining to minorities and women are pointed out.

Since this report is principally descriptive, no recommendations for increasing these low numbers are offered. The issue is articulated and some samples of intervention programs are described in Boxes 3.5 to 3.8. The data indicate clearly the seriousness of these circumstances and the consequences of continuing the current patterns.

1981, only 38 percent were fully proficient in computation at a sixth grade level, and 35 percent failed to demonstrate competence at this minimal level. Most discouraging of all, even the 7,000 students who had taken college preparatory courses in mathematics—algebra, geometry, and advanced algebra—did poorly. Only 4 percent of these were judged fully proficient in algebra and nearly two-thirds failed that portion of the test. Results for the test of verbal skills were hardly more encouraging. Of all the students who took the tests, 28 percent were rated as proficient, about 44 percent were lacking in one area (reading, vocabulary, grammar, writing) or another, and 28 percent failed in all areas.

Adequate preparation in mathematics in high school has been said to be the greatest single ticket to admission to and success in science and engineering careers. On the other hand, inadequate preparation in mathematic restricts major career choices and complicates an already difficult transition. There is evidence that this transition may cause many students to drop out of the pipeline toward mathematically based careers. One longitudinal study (1972 to 1979) showed that two of three students who were in the science and engineering pipeline at the end of the twelfth grade, that is, according to their planned college major, had dropped out of that pipeline by the junior year of college (NAS, 1987a). For blacks, the loss amounted to three of four students, and for Hispanics, seven of eight. Box 3.4 describes the general situation that exists concerning participation by minorities and women in the mathematical sciences. If mathematics is the ticket to success, then in an increasingly technological world, lack of it will be a stamp of exclusion. Efforts are being made to reverse this situation through intervention programs (see Boxes 3.5 to 3.8).

The dual nature of mathematics, which is both an academic competency and an academic subject, provides students with tools and concepts. The tools are necessary to capture a problem from any field in the proper quantitative terms; the concepts are what make mathematics an exciting discipline, and their mastery is a mark of an educated person. Thus "students need to be exposed to both faces of mathematics. They need to see that mathematics as an academic subject both depends on and strengthens mathematics as an academic competency; the content of the two aspects of mathematics should be in harmony" (CEEB, 1985a, p. 15).

Mathematics is a subject that builds on past knowledge, and further study in the field requires mastery of certain academic competencies. In addition to sharing forward logical progression with the other sciences, mathematics is unique in that once a student learns material in mathematics, it is assumed that that knowledge is retained forever. Thus it is very important that standards and expectations be uniform throughout the system and that there be no significant changes in the transitions from one type of school to another.

Remediation in College

The difficult transition from high school to college has affected students' attitudes about the role of college. In general, there appears to be a growing reliance on college to improve basic skills. More than 40% of entering freshmen in 1985 reported that an important reason for their attending college was to improve their reading ability and study skills; 70% said they were going to college to be able to make more money (CIRP, 1987b).

Feelings of being ill-prepared for college mathematics and reliance on remedial courses have been increasing among students. Almost one in ten (9%) entering freshman have already had some type of special tutoring or remedial work in mathematics. This percentage is much higher than that for any of the other fields; the next highest are for English (6%) and reading (5%). Additionally, a full one-fourth (25%) of all entering freshmen anticipated that they will need special tutoring or remedial help in mathematics, more than twice as many as those that predicted they will need help in English (12%), science (9%), reading (5%), or some other field. Although the same proportion of men as women reports past remediation or tutoring in mathemat-

---

BOX 3.5 Intervention Programs

Special programs in science, engineering, and mathematics are offered to encourage study by women and non-Asian minorities. The indications are that some of these intervention programs do work. Some of the key characteristics of academic-based intervention programs include the following (AAAS, 1984, p. 15):

- Academic component focused on enrichment rather than remediation;
- Highly competent teachers;
- Emphasis on applications and careers rather than on theory;
- Integrative approach to teaching;
- Multiyear involvement with students;
- Strong leadership;
- Stable, long-term funding base;
- Recruitment of participants;
- University, industry, and school cooperative program;
- Opportunities for in-school and out-of-school learning experiences;
- Parental involvement and community support;
- Specific attention to removing educational inequities;
- Development of peer support systems;
- Role models;
- Student commitment to "hard work";
- Evaluation, long-term follow-up, and careful data collection; and
- "Mainstreaming" of program elements into the institutional programs.

ics, more women (27%) than men (22%) anticipated that they will need further special tutoring or remedial help (CIRP, 1987b).

Not only do students anticipate that they will need remedial courses, but they do also in fact enroll in these courses. As many as one-fourth of all college freshmen are taking remedial courses in mathematics (BOC, 1988a). Enrollments in certain remedial courses—arithmetic, high school algebra and geometry, and general mathematics—have climbed steadily and steeply since the 1960s, much more so than enrollments in other mathematics courses (Figure 3.3). More students are needing and taking high-school-level mathematics courses in college, raising the much-discussed question: Should students expect to participate in higher education without the requisite background, and to what extent should colleges and universities try to accommodate these students? Those students capable of entering college should be provided with as much preparation in mathematics as possible before leaving high school, but many such students are not (CEEB, 1985a). The current wisdom is that all students should study mathematics on an academic track each year during high school (NRC, 1989).

Remedial courses, tutoring, and other supplements to normal college-level instruction became commonplace in colleges and universities during the period from 1970 to 1985. Today almost all two-year and four-year colleges offer remedial instruction or tutoring. In mathematics, remedial instruction increased dramatically from 1970 to 1985, with the expansion slowing down from 1980 to 1985. There are signs that the trend may be reversing, both in philosophy and practice.

In fall 1970, college enrollments in remedial courses constituted 33% of the mathematical sciences enrollments in two-year colleges and by 1985 had increased to 47%. In four-year colleges and universities, remedial enrollments constituted 9% of the mathematical sciences enrollments in 1970 and had increased to 15% by 1985. For fall 1985, these percentages translate to nearly three-fourths of a million enrollments—251,000 in four-year institutions and 482,000 in two-year institutions. The need for remedial instruction was ranked as the most serious problem in two-year mathematical sciences programs in the 1980 CBMS Survey and was still top-rated, along with the need to use temporary faculty for instruction, in the 1985 survey (CBMS, 1987). Departments in four-year colleges and universities rated remediation as a problem at a level that corresponded to the amount of remedial teaching required. Remediation was rated as a major problem by 39% of universities, by 66% of four-year public colleges, and by 45% of four-year private colleges. No responding statistics department rated remediation as a major problem (CBMS, 1987).

BOX 3.6 The Texas Prefreshman Engineering Program

The Texas Prefreshman Engineering Program (TexPREP) was started in 1986 as a statewide expansion of the successful San Antonio PREP program, begun in 1979 by Manuel P. Berriozabal. The purpose of TexPREP is to identify potential future scientists and engineers by identifying high school and middle school students of high ability and to provide these students with academic reenforcement to pursue science and engineering fields. The program operates at seven different locations throughout Texas.

Of the 2,000 students who have participated in TexPREP, more than three-quarters have been minority students and half have been women. Of those participants who are college-age, most (88%) plan to attend college or have graduated from college. A large share (68%) of TexPREP graduates major in science or engineering fields. TexPREP has a strong academic component, with courses in logic, algebra, engineering, computer science, physics, and technical writing. Other activities include field trips, guest speakers, and practice SAT examinations.

SOURCE: Information supplied by Manuel P. Berriozabal, University of Texas at San Antonio.

Remedial courses for most students are not considered to be preparatory for a mathematics-based college curriculum and are at best repeats of material usually taught in high school. There is little evidence of a significant effect on the supply of talented people in the pipeline aside from the knowledge and skills obtained through the content of these specific courses. Reversing an earlier pattern of low achievement in mathematics is too difficult for normal remedial programs. Mere repetition of material frequently results in duplication of previous failures to learn the necessary concepts (NRC, 1989).

Service Courses

In the 20 years from 1965 to 1985, total mathematical sciences enrollments in colleges and universities approximately doubled, increasing from 1.5 million per term to almost 3 million (Figure 3.4a). This included undergraduate and graduate enrollments in two-year and four-year institutions. Undergraduate mathematics enrollments in four-year colleges and universities increased by more than 60%, rising from about 1 million in 1965 to 1.6 million in 1985 (Figure 3.4c), while two-year college mathematics enrollments almost tripled, rising from 350,000 to slightly over 1 million (Figure 3.4d). In addition to remedial enrollments, there were large enrollments in service courses for other disciplines and in courses for general education. Approximately half of all enrollments were in courses below the level of calculus and half were in courses at or above that level. For all enrollments, including those in two-year colleges, about two-thirds were below the level of calculus and one-third were at or above that level.

The two-year college data included enrollments in mathematics, statistics, and some computing and data processing (Figure 3.4d). The large increases were almost all in remedial enrollments (accounting for nearly half the 1985 total) and in "other," which consists primarily of specialized vocational courses, many of which are called technical mathematics. These courses generally have a low-level content, some being arithmetic-based and some being algebra-based.

The increases in undergraduate mathematics enroll-

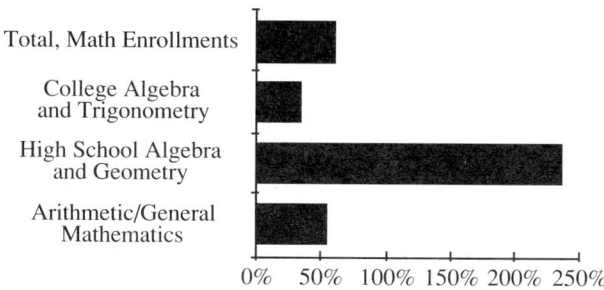

FIGURE 3.3 Percent increase in enrollments in selected mathematics courses in colleges and universities, 1965 to 1985. SOURCE: Conference Board of the Mathematical Sciences (CBMS, 1987).

ments in four-year colleges and universities were in remedial, precalculus, and calculus-level courses. The calculus enrollments shown in Figure 3.4 include those in differential equations and in linear algebra. Advanced course enrollments, those above the level of calculus, have increased only slightly, with much of the increase having occurred since 1980.

Most students who study the mathematical sciences in colleges and universities for general educational purposes enroll in courses that are designed to serve particular needs in various curricula that use mathematics. Among these courses are ones in college algebra and trigonometry, finite mathematics, calculus, statistics, and liberal arts mathematics. Only the liberal arts mathematics course, and possibly the statistics course, are likely to have been designed with the general education of the student in mind.

Enrollments in liberal arts mathematics courses peaked in 1975 at 175,000 and have dropped dramatically since then, to 70,000 in 1985 (CBMS, 1987). Enrollments in elementary probability and statistics courses increased significantly from 126,000 in 1975 to 180,000 in 1985. Additionally, computer science courses became generally available and popular in the 1970s. Since many such courses have no college mathematics course as a prerequisite, these are likely alternatives to mathematics courses for general education. Enrollments in elementary computer science courses were estimated at more than 250,000 in fall 1980 and at over 400,000 in fall 1985 (CBMS, 1987).

Mathematical sciences departments provide "hard" service courses for many college curricula, most notably in the traditional areas of the physical sciences and engineering. "Hard" service courses are those with specified content that will be needed in students' later studies, as opposed to "soft" courses, which have few or no restrictions on content. "Hard" service courses with large enrollments have become much more common in recent years for students in business, the social sciences, the life sciences, and preprofessional curricula.

One critical service course area provided by mathematical science departments is for students in teacher education programs. These students include prospective secondary school mathematics teachers (see Chapters 4 and 5) and prospective elementary school teachers. The courses that prospective secondary school mathematics teachers take range from college algebra through advanced undergraduate courses; requirements vary widely across the country. Prospective elementary school teachers are likely to take one or two mathematics courses especially designed for them, but most do not take any other college-level mathematics courses (OTA, 1988b, p. 65).

Survey data on courses taken by college students from 1980 to 1984 indicate that an average college graduate takes 8.4 semester hours of mathematics, which translates to between two and three mathematics courses in college for the average bachelor's degree holder. The range includes a high (for nonmathematics majors) of 21 hours (or about six to seven courses) for computer science majors and a low of 2 hours (less than one course) for fine arts and English majors. Although definitive data are not available, there are indications that the amount of mathematics studied by college students has increased in the past decade (see Appendix Table A3.7).

The fraction of college students who take mathematics courses and their success in those courses compared to their other courses give an indication of the difficulty students have with mathematics. To determine course-taking patterns, the Department of Education conducted a national longitudinal study based on an analysis of the college transcripts of 1972 high school graduates who attended college. A significant share, two of five, took no mathematics courses at all in college, and another one of five took only one course, that is, one to three credits in mathematics. (Even though these data are old, more recent data are neither available nor expected until 1992.) Students found mathematics courses difficult as evidenced by lower grade point averages (GPAs) in mathematics courses

---

BOX 3.7 Professional Development Program

The Professional Development Program (PDP) at the University of California, Berkeley, houses the Charles A. Dana Center for Innovation in Mathematics and Science Education, which is directed by Philip Uri Treisman. The Dana Center program promotes achievement in mathematics courses for minorities by providing an environment where students can learn and by fostering productive study habits. The study group approach, modeled on Asian study groups, incorporates many of the key characteristics of successful intervention programs: high expectations of competence, a strong academic component, capable and appropriate instruction, cooperative learning, and commitment from students. A critical feature is an assumption of competence, the Dana Center program being regarded as an honors program as opposed to a remedial one.

This program has been associated not only with successful completion of calculus courses by more of the minority students who participate, but also with high retention and graduation rates. The program has been expanded to include other California universities, and Treisman is currently working on a high school program in mathematics for minorities.

SOURCE: AMS Notices, "Research Mathematics in Mathematics Education," Volume 35, Number 8, October 1988, American Mathematical Society, Providence, R.I., pp. 1123-1131.

than in other courses. Only about 10% had an overall GPA of less than 2 (on a scale of 4), but for mathematics courses 35% had a GPA of less than 2; half had overall GPAs between 2 and 3, but only one-third fell in this range for mathematics courses; and 35% of students in the sample had GPAs at the top end of the range, between 3 and 4, but only 29% fell within this range for mathematics courses.

Mathematics as an Academic Competency and Subject

As stated above, mathematics is both an academic competency and an academic subject. In recent years, the demand for mathematics as a competency—in service courses—has increased dramatically. At the same time, the number of students choosing mathematics as a major has decreased, creating, among other things, an increasing demand for mathematics teaching and a decreasing supply of mathematics teachers. Chapters 4 and 5 give data on majors in mathematics and statistics and on the utilization of these majors in the workplace. Teachers, for both school and college, will be a principal subject because of the increasing demand for mathematics education.

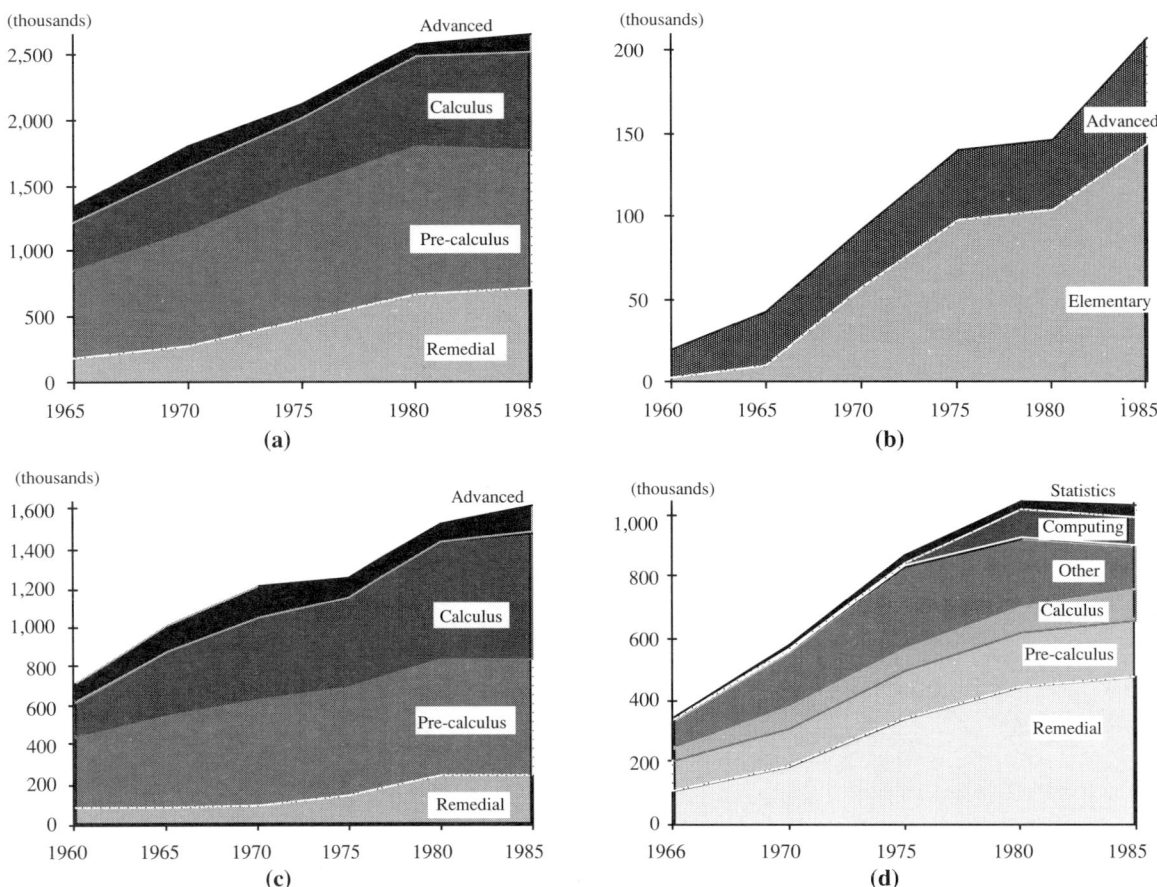

FIGURE 3.4 Undergraduate enrollments in mathematical sciences departments at U.S. colleges and universities. (a) Total, 1965 to 1985; (b) in statistics, 1960 to 1985; (c) in four-year colleges and universities, 1960 to 1985; and (d) in two-year colleges, 1966 to 1985. SOURCE: Conference Board of the Mathematical Sciences (CBMS, 1987).

BOX 3.8 The Mathematics, Engineering, Science Achievement Program

The Mathematics, Engineering, Science Achievement (MESA) Program was established with the goal of increasing the number of black, Hispanic, and Native American students completing bachelor's degrees in California in the fields of mathematics, science, or engineering. Begun in 1970, the program is based at Lawrence Hall of Science in Berkeley, California, and operates under the auspices of the University of California at Berkeley. Because of its success in recruiting and training minority students at the junior high, high school, and undergraduate levels for science and engineering degrees, the MESA program in California has served as a model for other states.

Internships, field trips, incentive awards, counseling, freshman orientation and guidance, financial aid and scholarships, and student study groups are some of the activities provided by the program. Students are encouraged through MESA's Pre-College Program to take preparatory classes in mathematics and science in junior and senior high school. These courses, although usually optional for students, are critical to their remaining in the science and engineering pipeline. Most of the high school graduates participating in MESA have pursued mathematics-based majors. The retention rates in college of MESA participants are considerably higher than those for nonparticipants.

SOURCE: Office of Technology Assessment, Educating Scientists and Engineers: Grade School to Grad School, p. 39 (OTA, 1988a).

4 Majors in Mathematics and Statistics

- Attrition from the mathematical sciences pipeline is approximately half per year after the ninth grade.

- The number of degrees awarded annually at each of the three degree levels increased sharply after 1960, peaked in 1970, and then declined sharply to about the mid-1960s level and about the average levels of the past 40 years.

- Many students decide to major in mathematics after entering college, and the circumstances in graduate and undergraduate programs are closely connected, being simultaneously subject to the same forces.

- Relatively few women, blacks, and Hispanics receive degrees in the mathematical sciences, especially at the graduate level.

- Graduate students, nearly one-half of whom are not U.S. citizens, are mostly supported by teaching assistantships.

Introduction

The prebaccalaureate "pipeline" population in mathematics and statistics is elusive as students move in and out of degree programs. The attrition rate for this population is certainly not uniform from year to year, but, for recent years, assuming an attrition rate of 50% per year from the ninth grade onward gives surprisingly close estimates of the actual number of bachelor's, master's, and doctoral degrees awarded in mathematics and statistics (Figure 4.1).

As indicated in Figure 4.1, in 1972 there were approximately 3.6 million U.S. students in the ninth grade. An attrition rate of 50% per year yields about 225,000 college freshmen in 1976. An attrition of 50% in each of the next four years of college yields about 14,000, approximately the number of bachelor's degrees awarded in mathematics and statistics in 1980. (The actual number reported for 1980 was 11,000 (NCES, 1988a), but this figure was near the minimum reported since 1960 and has increased since.) Halving twice more on the way to the master's degree yields 3,500 degrees (there were 2,700 reported in 1982), and three more halvings yield 437 doctoral degrees in 1985, which closely approximates the 400 U.S. citizens who received doctorates in 1986. (It is noted that the assumption of five years from the baccalaureate to the doctorate in the model is short of the average of six years of registration, and there are no adjustments for non-U.S. students below the doctoral level.)

The losses of mathematical talent are not evenly distributed among racial and ethnic groups or the sexes. At each critical juncture in the pipeline more women and minority students drop out than do white males. In the eighth grade the mix of students is roughly equal to their representation in the population. Fifteen years later this representation is highly skewed in favor of non-Hispanic white males, with 78% of doctoral degrees being earned by 42% of the population (Figure 4.2). The highest losses in the pipeline occur among blacks and Hispanics, two segments of the population that are increasing.

A Challenge of Numbers

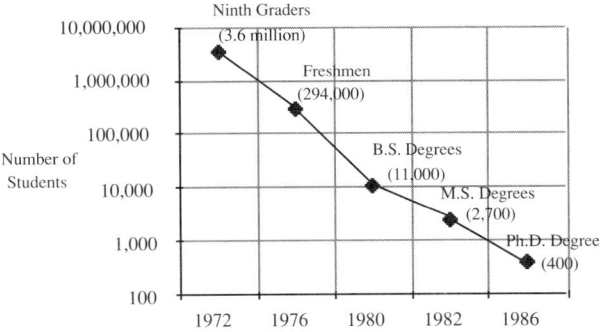

FIGURE 4.1 Students in the mathematical sciences pipeline—about half are lost each year. SOURCES: National Center for Education Statistics (NCES, 1987a), Cooperative Institutional Research Program (CIRP, 1987b), and American Mathematical Society (AMS, 1986 to 1988).

The numbers of mathematical sciences degrees awarded annually to U.S. citizens and others at the bachelor's, master's, and doctoral levels have shown similar patterns since 1950 (Figure 4.3). Steep increases occurred from the mid-1950s to peak production in the early 1970s. Equally steep decreases then occurred until the early 1980s, and moderate increases have occurred since. Although similar patterns exist in other areas of science and engineering, the numbers of mathematical sciences degrees have been the slowest to rebound from the declines of the 1970s and early 1980s, and degree production is still very low even when compared to that for the other sciences and for engineering. Current numbers of degrees awarded annually are approximately the same as those for the early 1960s and very near the averages for the past 40 years: 15,000 bachelor's degrees, 3,000 master's degrees, and 800 doctoral degrees. These numbers are each 40-45% lower than the analogous numbers for peak production in the early 1970s. This supply appears to be slightly short of current demand, and significantly higher demands at all three levels are projected.

The composition of the population of degree recipients has changed over the past 15 years. The most dramatic change has been in the new-doctoral-degree population, in which the fraction of U.S. citizens has tumbled from four-fifths to less than one-half. The other significant change has been in the representation of women among degree holders, particularly at the baccalaureate level. The percentage of women among the annual recipients of bachelor's degrees has increased steadily so that current levels almost represent parity. This is not the case, however, at the master's degree and doctoral degree levels.

Recent increases in the number of bachelor's degrees appear to be fueled by students changing to a major in mathematics after entering college, since the number of entering freshmen expressing an intent to major in mathematics remains very low, at less than 1%. Consequently, projections of future numbers of degrees based on the current number of planned majors are tentative. That current num-

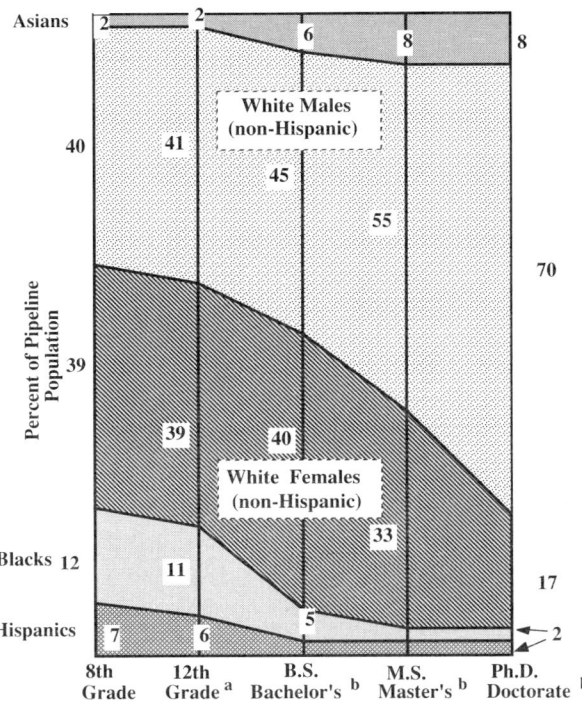

[a] High school record indicates possible choice of mathematics as a college major
[b] Degrees in mathematics

FIGURE 4.2 A representation of U.S. students in the mathematics pipeline. SOURCES: Bureau of the Census (BOC, 1982 and 1986), National Center for Education Statistics (NCES, 1987a), and American Mathematical Society (AMS, 1986 to 1988).

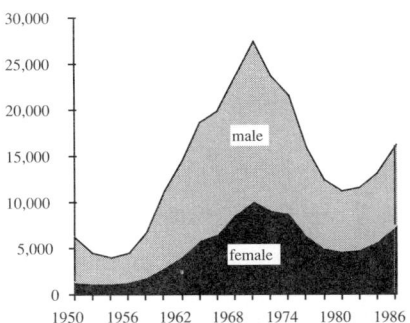 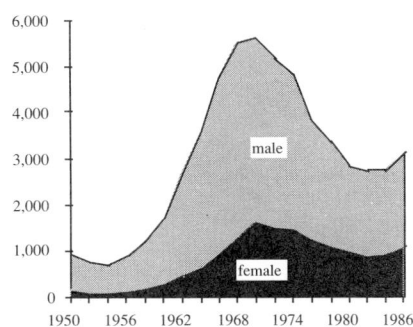 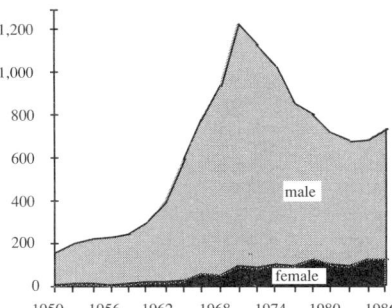

FIGURE 4.3 Number of mathematical sciences degrees awarded by U.S. institutions, 1950 to 1986. Left: Bachelor's degrees. Middle: Master's degrees. Right: Doctoral degrees. SOURCE: National Center for Education Statistics (NCES, 1988a).

ber does not predict significant increases in degree production in the near future (Figure 4.4).

There is also a striking similarity in the trends in the numbers of degrees awarded at the three levels in that no noticeable phase lag exists from bachelor's to master's to doctoral degrees. This suggests that the attitudes about majoring in mathematics originate in the colleges and universities and that the undergraduate and graduate programs are simultaneously subject to the same forces.

Undergraduate Majors

The number of baccalaureate degrees awarded in mathematics rose from 11,000 in 1960 to a high of 27,000 in 1970, declined steadily throughout the 1970s and early 1980s, and by 1986 had rebounded slightly to the current level of 16,000 (Figure 4.5). The fraction of freshmen anticipating a mathematical sciences major has dropped sharply, from 3.3% in 1970 to 0.6% in 1987, and the share of degrees awarded to women has increased steadily, from about one-third in the mid-1960s to almost one-half (46%) in 1986.

Over the past 20 years total mathematical sciences enrollments and the number of mathematical sciences majors have reflected disparate trends. For example, the number of bachelor's degrees awarded per 1,000 mathematical sciences course enrollments in four-year institutions has varied from 7 to 23, and from 4 to 15 if two-year college enrollments are included (Table 4.1).

At the undergraduate level there have been close ties between the mathematical sciences and computer science (see Box 1.1). The decline in the number of degrees in mathematical sciences since 1970 has occurred at the same time as a comparable increase in the number of degrees in computer science (Figure 4.6). Joint majors in mathematics and computer science and overlapping employment opportunities are additional indications of the interconnections.

When or why students decide to major in mathematics is not clear, although evidence suggests that many make

TABLE 4.1 Mathematical sciences bachelor's degrees per 1,000 mathematical sciences enrollments, 1965 to 1985

	1965	1970	1975	1980	1985
Four-year enrollments only	20	23	15	7	9
All enrollments	15	15	9	4	6

SOURCES: Adapted from Conference Board of the Mathematical Sciences (CBMS, 1987) and National Center for Education Statistics (NCES, 1987b).

these decisions after entering college. Comparing the number of full-time freshmen who anticipated a mathematics major with the actual number of mathematics degrees conferred four years later shows clearly that recently many students have decided to major in mathematics after entering college. The expected number of bachelor's degrees in the mathematical sciences is correlated with freshmen interest, and decreases at the same rate as that level of interest, since freshmen enrollments have remained relatively stable.

Several studies have shown, however, that broad decisions about science versus nonscience careers are made by the time a student graduates from high school, and high achievement in mathematics at the secondary school level is predictive of a mathematics-related career (NAS, 1987a). Knowledge of mathematics has been said to be central in separating scientists from nonscientists, and its significance is increasing in areas of business and management. Major losses, as high as 50%, take place during the college years in the general science and engineering talent pipeline, but recently it appears that college students have been slowly migrating toward, rather than away from, mathematics as a field of study.

Prior to 1975 many more entering students planned a mathematics major than actually received a mathematics degree. But in each year since the early 1980s the number

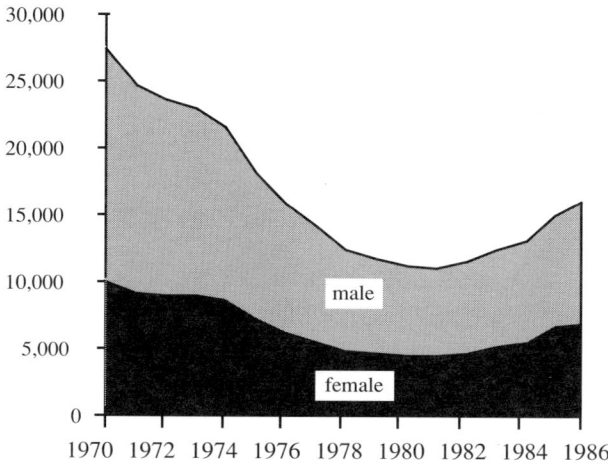

FIGURE 4.5 Bachelor's degrees awarded in the mathematical sciences, 1970 to 1986. SOURCE: National Center for Education Statistics (NCES, 1988a).

of students who have entered college intending to major in mathematics has actually been lower than the number receiving a bachelor's degree in mathematics four years later. The interest of entering freshmen in mathematics as a probable major has been very low in the 1980s, but degree production has increased (Figure 4.7). Thus the anticipated major of college freshmen as measured by the Cooperative Institutional Research Program's survey (CIRP, 1987b) is not a very good forecast of the number of mathematics degrees awarded four years later. The discrepancy between expected and actual major and the recent changing migration of talent into rather than out of the mathematics pipeline make any discussion of persistence in the choice of major somewhat speculative.

High achievers, as would be expected, are more likely to persist in their freshman choice of a mathematics major. A 1985 study of students who were college freshmen in 1981 showed that most freshmen who had planned to major in mathematics and had achieved an "A" grade point average after four years of college did in fact major in mathematics. Students with lower grade point averages were less likely to continue with their freshman choice of a mathematics major and to switch instead to other fields, most often to the social sciences or business (Table 4.2).

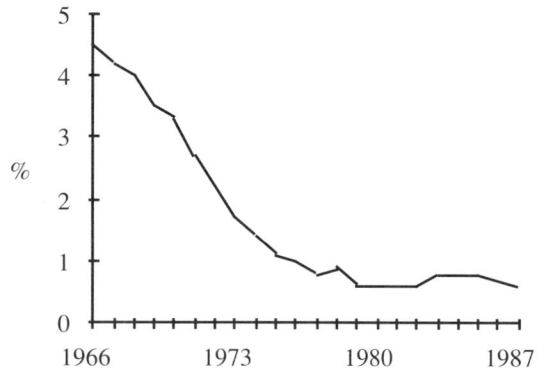

FIGURE 4.4 Percentage of entering college freshmen expecting to major in mathematics. SOURCE: Cooperative Institutional Research Program (CIRP, 1987b).

Majors in Mathematics and Statistics

One notable trend, mentioned above, is the increased representation of women among recipients of baccalaureate degrees in mathematics. In 1950 less than one-fourth of the 6,000 mathematics degrees were awarded to women, and in 1986 almost half of 16,000 were. This progress in the representation of women at the baccalaureate level, however, has not translated to corresponding progress at the graduate level. Even though the fraction of women among baccalaureate degree recipients has been at more than 40% since 1975, only 27% of the current full-time graduate students enrolled in mathematical sciences doctorate-granting institutions are women. This is a priority issue of the Association of Women in Mathematics: "The widespread and successful participation of women in undergraduate mathematics must be better understood in order to counteract the attrition that occurs at other stages of the educational process, most notably in adolescence and in the graduate years" (AWM, 1988, p. 9).

Blacks and Hispanics receive few of the bachelor degrees awarded in the mathematical sciences. The ethnic and racial composition of U.S. citizens receiving bachelor's degrees in 1985 in the mathematical sciences (and in all fields, for comparison) is given in Table 4.3. Non-U.S. citizens received only a small fraction (5%) of these degrees.

In 1982, according to an opinion survey, most senior college and university academic administrators (not departmental chairs) did not think that the quality of undergraduate science and engineering students had declined during the previous five years. The majority felt that there had been no change in quality, and one-fourth thought there had been improvement. These administrators also thought that the most able students who were shifting fields were shifting toward rather than away from science and engineering (ACE, 1984). However, a more recent survey of mathematical sciences departments regarding the quality of undergraduate majors in mathematics indicates that lack of quality is a major concern.

The 1985-1986 CBMS Survey asked mathematical sciences departments to rate the lack of quality and the lack of quantity of undergraduate majors as problems in their programs (CBMS, 1987). About half of the responding departments rated the lack of quantity as a major problem, and a slightly higher percentage rated lack of quality as a major problem (Table 4.4).

Undergraduate mathematics majors are the primary source of students for graduate programs in the mathematical sciences. As is true for doctoral degree holders in the physical sciences and engineering, almost three-quarters of those who receive doctorates in mathematics also have

TABLE 4.2 Changes in mathematical sciences majors by undergraduate grade point averages, 1981 freshman cohort

	Undergraduate grade point average[a]		
	A	A- or B+	B or less
Percent persisting in freshman choice	87	22	5
Percent defecting to			
Biology	N	N	N
Physical sciences	13	N	3
Social sciences	N	28	14
Engineering	N	N	20
Business	N	19	14
Education	N	8	12
Technical fields	N	5	15
Other	N	19	17

[a] N means no significant number.
SOURCE: Holmstrom, E. I., unpublished data.

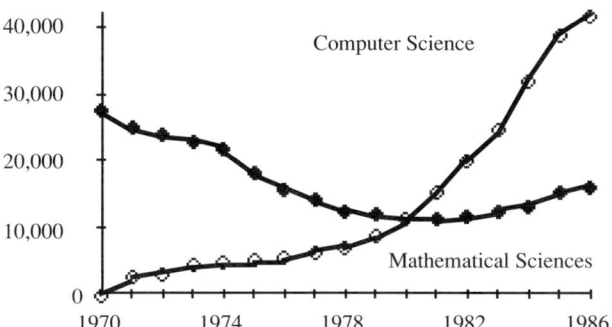

FIGURE 4.6 Number of bachelor's degrees awarded in mathematical and computer sciences, 1970 to 1986. SOURCE: National Center for Education Statistics (NCES, 1988a).

39

a bachelor's degree in the same field. Nearly 20% of recent mathematics graduates have enrolled as full-time graduate students, but often in areas other than mathematics or statistics (see section headed "Doctoral Degree Recipients"). Those not continuing their studies have found favorable employment opportunities.

In 1986 a low general unemployment rate (2%) and a high science and engineering employment rate (74%) characterized the labor market for recent mathematics bachelor's degree recipients. This compared with rates for bachelor's degree holders in all fields of 4% and 64% for general unemployment and for science and engineering employment, respectively. Industry employed more than half of recent mathematics graduates, one-fourth found teaching positions, and the remainder worked for government. On average, of every five mathematics graduates employed in a science or engineering field, two found work in a mathematics or statistics field, two in computer science, and one in engineering, economics, or some other field. The 1986 median annual salary of $24,100 for mathematics bachelor's degree holders was just below the average of $25,000 for all science and engineering fields, trailed the average for engineering ($30,000) and for computer science ($28,000), but topped that for all other science fields. Salaries were higher by about $7,000 in industry and government than they were in educational institutions.

Degrees for Secondary School Mathematics Teachers

Two different but related trends are primarily responsible for the current low number of degrees being awarded to prospective secondary school mathematics teachers. Interest in the study of both mathematics and education declined sharply in the 1970s and early 1980s. This drop in popularity has translated to fewer degrees in both fields (Figure 4.8). The following is a discussion of the numbers and some characteristics of persons receiving degrees as preparation for secondary mathematics teaching. A related discussion of school mathematics teachers in the workplace is in Chapter 5.

The loss of interest in teaching degrees dipped to its

TABLE 4.3 1985 bachelor's degrees awarded in mathematical sciences

	U.S. Citizens							
	White	Black	Hispanic	Asian	Indian	Total U.S.	Non-U.S.	Total
Total, math. sci.	12,162	766	257	880	59	14,124	761	14,885
Percent								
By race in the U.S.	86.1%	5.4%	1.8%	6.2%	0.4%	100%		
By citizenship						95%	5%	100%
Men								54%
Women								46%
Total, all fields								979,477
Percent								
By race in the U.S.	88.0%	6.1%	2.8%	2.7%	0.4%	100%		
By citizenship						97%	3%	100%
Men								49%
Women								51%

SOURCE: NCES, unpublished data.

TABLE 4.4 Summary of responses on quality and quantity of undergraduate majors

	Mathematics			Statistics
	Universities	Public Four Year	Private Four Year	Universities
Lack of quality	38% major	62% major	39% major	31% major
	15% no/minor	6% no/minor	7% no/minor	9% no/minor
Lack of quantity	39% major	54% major	42% major	22% major
	18% no/minor	20% no/minor	9% no/minor	21% no/minor

NOTE: The possible responses were 0, 1, 2, 3, 4, or 5 with 0 meaning "no problem" and 5 meaning "major problem," and the others indicating gradations between these. The percentages given above for "major" represent the 4 and 5 responses, while "no/minor" represents the 0 and 1 responses. By subtracting the sum of these percentages from 100, one can get the percentage of 2 and 3 responses.
SOURCE: Conference Board of the Mathematical Sciences (CBMS, 1987).

lowest level in 1982 and has rebounded slightly since. The reasons for the loss of appeal of teaching are varied and complex, but they include low salaries, broader career options for women, and lack of prestige associated with teaching (OTA, 1988b). Additionally, market conditions for teachers throughout the 1970s were not favorable because of a perceived general oversupply that actually was present in some disciplines but not necessarily in mathematics. The recent turnaround is supported by increased public awareness, higher salaries, and a more favorable market. However, increases are not expected to be sufficient to meet anticipated demand.

The number of baccalaureates conferred in education and the number in mathematics have declined as dramatically as the interest levels of entering freshmen. From 1971 to 1985, education degrees awarded annually decreased by 50%, and mathematics degrees decreased by 39%. Fifteen years ago, one in five baccalaureate recipients majored in education; today, the number is less than one in ten.

Most new teachers were education majors in college, but some were single-subject (such as mathematics) majors who did supplementary work in education. The single-subject majors appear to be increasing. This switch from education to subject majors reflects the current questioning of the usefulness of an education degree: "The utility of the education major is under serious consideration at the moment and several groups have proposed a wide-ranging overhaul of teacher education" (OTA, 1988b, p. 54).

Only a small portion of those majoring in education have specialized in mathematics education and hence are likely to become high school mathematics teachers. About 1% of the bachelor's degree holders in education, 0.5% of the master's degree holders, and 0.4% of the doctoral

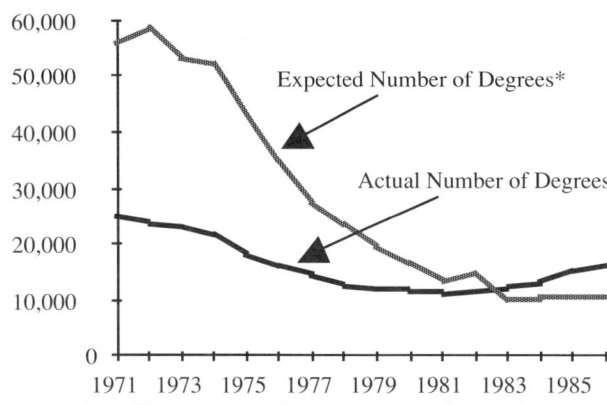

* Number of freshmen entering four years earlier and intending to major in mathematics.

FIGURE 4.7 Expected versus actual number of bachelor's degrees in mathematical sciences. SOURCES: Cooperative Institutional Research Program (CIRP, 1987b) and National Center for Education Statistics (NCES, 1987a).

degree holders have specialized in mathematics. Since the early 1980s, education departments have produced between 1,000 and 1,500 candidates for high school mathematics teaching each year. Recently, there has been an upswing in the number of bachelor's degrees awarded in mathematics education (see Appendix Table A4.4); however, these figures are small when compared to the 16,000 school districts that need mathematics teachers.

In addition to education majors specializing in mathematics, the other main source of high school mathematics teachers is undergraduate mathematics majors. A survey of recent mathematics graduates conducted by the NSF in 1986 found that about one-quarter of bachelor's degree recipients taught in educational institutions, presumably high schools, and one-third of master's degree holders also taught, some in high schools and others in college (NSF, 1987a).

Standardized test scores for students in the field of education are lower, on the average, than are those for students in other general areas of study, but data on the relative qualifications and abilities of students attracted to education are limited. The SAT mathematics and verbal scores for those planning education majors are lower than the average scores for all SAT takers. But no breakdown of these scores is available by intended level or field of teaching, and so it is not possible to determine how those planning to teach secondary school mathematics compare

TABLE 4.5 1987 SAT scores by intended college major

	Mathematics	Verbal
Business and commerce	459	408
Education[a]	437	408
Engineering	554	456
Language and literature	518	537
Mathematics	602	475
Physical sciences	576	507
TOTAL	476	430

[a] Includes elementary and secondary education majors.
SOURCE: College Entrance Examination Board (CEEB, 1987).

TABLE 4.6 1986 GRE scores by undergraduate and intended graduate major

	Undergraduate major		Intended graduate major	
	Quant.	Verbal	Quant.	Verbal
Education[a]	451	441	464	457
Engineering	674	464	671	461
Math. sciences	657	484	657	481
Physical sciences	637	505	639	504
Other humanities	517	543	523	542
Total	540	489	540	489

[a] Includes elementary and secondary education majors.
SOURCE: Educational Testing Service (ETS, 1987).

with others, both in education and mathematics. The 1987 SAT scores by intended major for selected fields are given in Table 4.5.

Graduate Record Examination (GRE) scores for 1986 for students planning graduate work show that students who majored in education scored below the national averages, almost 90 points on the quantitative portion and 47 points on the verbal portion (Table 4.6). Those who planned graduate work in education also had below-average scores in both the quantitative and verbal components of the GRE.

The above discussion reflects a serious lack of information about students preparing for careers in teaching. There is little information that distinguishes entrants into the various teaching fields, and there is a wide variety of degree programs indicative of the differing certification requirements across the country.

Graduate Students

Both the quantity and quality of graduate students are considered problems by most mathematical sciences graduate departments. In 1986 most of the approximately 18,000 mathematical sciences graduate students were enrolled in doctorate-granting institutions. Over 40% of those enrolled full-time were non-U.S. citizens (Figures

4.9 and 4.10), and that fraction was greater in the top graduate (Group I) institutions (Figure 4.11). Fewer women, blacks, and Hispanics enroll in graduate programs than their share among bachelor's degree holders would predict. The dominant mode of support is graduate teaching assistantships, with the result that nearly one-half million students—one of twelve—each year are taught solely by graduate assistants. Three-fourths of the graduate students have undergraduate degrees in mathematics.

Of the 18,000 graduate students enrolled in mathematical sciences programs in 1986, almost 12,000 were enrolled full-time in doctorate-granting institutions, and the remainder were either part-time students or were in master's-granting institutions. From 1975 to 1986 the total number of students enrolled full-time increased by about 1,700, or 17%. This total gain of 1,700 students actually represents a loss of U.S. students (1,400) and a gain of non-U.S. students (3,100). The number of U.S. students enrolled reached a low in 1981 and has increased since to the current level. The number of non-U.S. students has almost tripled. The change in the composition of graduate students has been significant in the mathematical sciences, even when compared to other science and engineering fields experiencing the same general trends (Figure 4.9).

In the mathematical sciences the number of U.S. students enrolled was actually less in 1986 than in 1975 (Figure 4.10). This drop in U.S. students combined with the increased numbers of non-U.S. students has had an impact on both teaching and learning. Since most graduate students in the mathematical sciences are teaching assistants, this increasing reliance on non-U.S. students has affected undergraduate teaching and is believed to be affecting the choice of mathematics or mathematics-related majors.

Because undergraduate education in many countries is much more specialized than it is in the United States, non-U.S. students have often been exposed to much more mathematics and thus come to graduate school with a more sophisticated understanding of the field than do their U.S. counterparts. For instance, in China a baccalaureate degree in mathematical sciences is comparable to the traditional master's degree in the United States (MAA, 1983). Such countries have opted for depth of specialization over the breadth of a liberal arts education.

Besides the decreasing number and proportion of Americans, the other striking characteristic of mathemati-

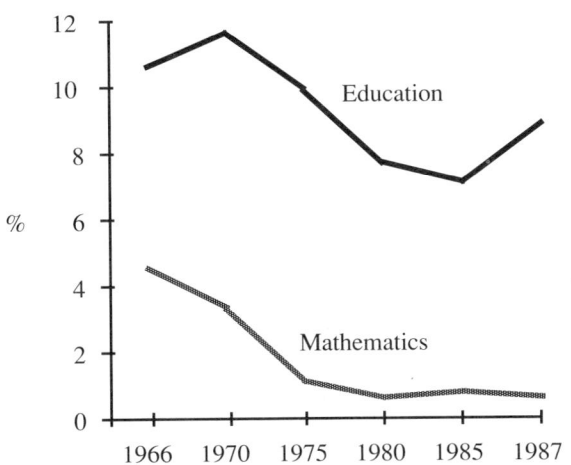
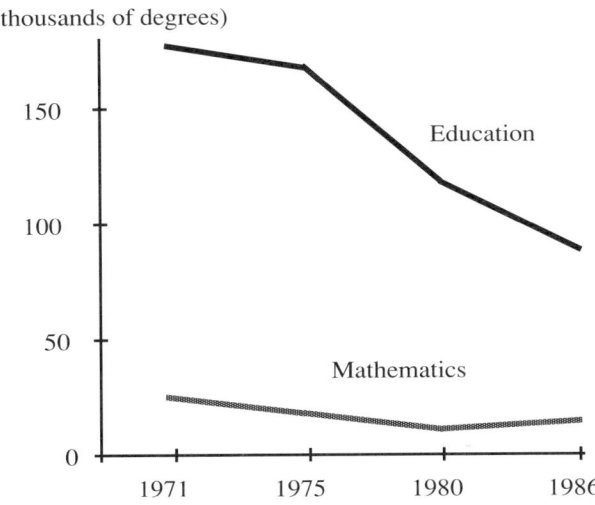

FIGURE 4.8 Left: Interest in matnematics and education among entering college freshmen. Right: Degrees in mathematics and education among exiting college seniors. **SOURCES: Cooperative Institutional Research Program (CIRP, 1987b) and National Center for Education Statistics (NCES, 1987a).**

cal sciences graduate students in doctorate-granting institutions is that almost three-quarters are men. However, there are about 1,000 more women currently enrolled full-time in doctorate-granting institutions than there were in 1975, and the fraction of women has increased from 21% in 1975 to 27% in 1986 (see Appendix Table A4.6). Small numbers of non-Asian minority students are enrolled in graduate programs; however, in terms of percentages, more such students are enrolled in master's-granting institutions than are enrolled in doctorate-granting institutions (see Appendix Table A4.9).

Mathematical sciences graduate students depend heavily on institutional teaching assistantships for support during their studies. In 1986, 70% listed institutional support as their major source of income, 17% were self-supporting, 8% reported having federal support, and 5% listed other outside support (Figure 4.12). There were gender differences in the sources of major support for full-time graduate students in doctorate-granting institutions. Women were more likely to have institutional and self-support than were men, and they were less likely to receive federal and other outside support than were the male graduate students.

The various types of support for graduate study include fellowships, traineeships, research assistantships, teaching assistantships, and other types of support. Of the graduate students in the mathematical sciences in 1986, 59% supported themselves with teaching assistantships, 9% had research assistantships, 7% had fellowships, 1% had traineeships, and 24% reported other types of support. Nearly half the teaching assistants taught their own classes, while the others conducted recitations, tutored, and graded papers. Compared to physical sciences and engineering graduate students, mathematical sciences graduate students are much more likely to teach and much less likely to have research assistantships. One-quarter of all science and engineering graduate students have research assistantships, and despite increases in recent years, only 9% of those in mathematics do. Although about one-quarter of all science and engineering students have teaching assistantships, fully three-

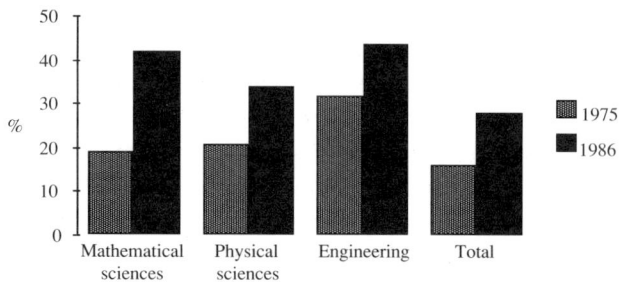

FIGURE 4.9 Percent of full-time graduate students in doctorate-granting institutions who are non-U.S. citizens, 1975 and 1986. SOURCE: National Science Foundation (NSF, 1988a).

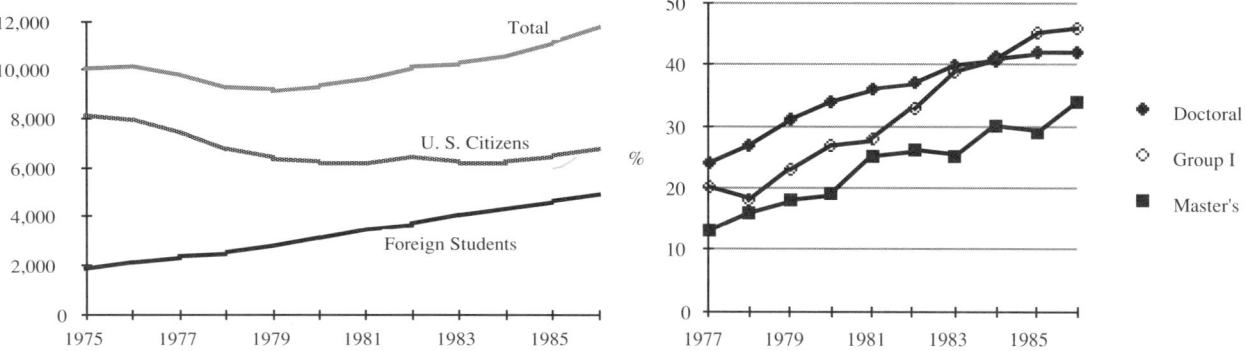

FIGURE 4.10 (Left) Mathematical sciences graduate students enrolled full-time in doctorate-granting institutions, 1975 to 1986. (See Appendix Table A4.5.) **FIGURE 4.11** (Right) Percent of non-U.S. citizens as mathematical sciences graduate students by type of institution, 1977 to 1986. (See Box 3.2 or note on Appendix Table A5.13 for explanation of Group I.) SOURCES: National Science Foundation (NSF, 1988a) and Conference Board of the Mathematical Sciences (CBMS, 1987).

TABLE 4.7 Summary of responses on quality and quantity of graduate students

	Mathematics		Statistics
	Universities	Public four-year colleges	Universities
Lack of quality	50% major	44% major	56% major
	2% no/minor	21% no/minor	14% no/minor
Lack of quantity	52% major	53% major	55% major
	14% no/minor	24% no/minor	20% no/minor

NOTE: The possible responses were 0, 1, 2, 3, 4, or 5 with 0 meaning "no problem" and 5 meaning "major problem," and the others indicating gradations between these. The percentages given above for "major" represent the 4 and 5 responses, while "no/minor" represents the 0 and 1 responses. By subtracting the sum of these percentages from 100, one can get the percentage of 2 and 3 responses.
SOURCE: Conference Board of the Mathematical Sciences (CBMS, 1987).

fifths of those in mathematics do. Some of the differences in support among fields are due to the nature of the field, and particularly to whether laboratory work is significant or not—it has not been significant in the mathematical sciences. "Other types of support," a category that includes self-support, was highest among engineers at 38%, lowest among physical scientists at 9%, and in between among mathematical sciences graduate students at 24% (Figure 4.13). The percent of graduate students with fellowships was about the same in the mathematical sciences, the physical sciences, and engineering.

Almost three-quarters (74%) of prospective graduate students who take the GRE and state an intention to do doctoral work in mathematical sciences have an undergraduate major in mathematics (ETS, 1987). Another 22% have majored in another science or engineering field, and 4% are nonscience majors. Among those intending to do graduate work below the doctoral level, the large majority (68%) have a baccalaureate degree in mathematics, 26% have majored in another science or engineering field, and 6% have a baccalaureate degree in a nonscience field.

Even though most students who pursue graduate-level work in the mathematical sciences have backgrounds that include degrees in the same field, the quality of incoming students is often not what graduate programs expect. The 1985-1986 CBMS Survey asked responding mathematical sciences departments to rate the lack of quality and the lack of quantity of graduate students as problems in the programs (CBMS, 1987). More than half of the responding graduate departments rated both the lack of quantity and the lack of quality as major problems (Table 4.7).

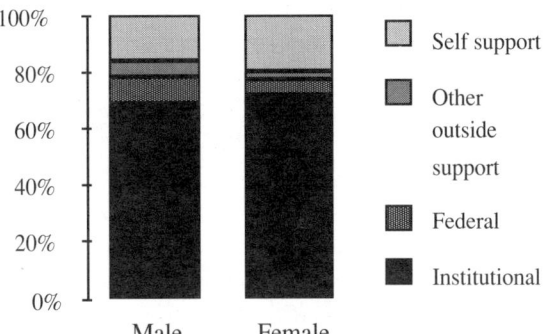

FIGURE 4.12 Source of major support for mathematical sciences graduate students in doctorate-granting institutions, 1986. (See Appendix Table A4.7.) SOURCE: National Science Foundation (NSF, 1988a).

Master's Degree Recipients

The production of mathematical sciences master's degrees has followed the same pattern as that for both bachelor's and doctoral degrees, peaking in 1970 after a steady climb and dropping since then. The number of degrees steadily increased from about 1,000 in 1950, peaked at 5,600 in 1970, and has decreased since with a slight rebound in the last few years. In 1986 approximately 3,200 master's degrees were awarded. The current level of production is the same as that in the mid-1960s, represents a drop of 44% from the peak production of 1970, and is about equal to the average for the past 40 years (Figure 4.14).

Perhaps what is most notable about recipients of master's degrees is just how little is written or known about them. There are several major surveys that track doctoral degree holders, but these include few details on master's degree recipients. Little is known about how master's degree holders fit into the broader mathematical community. Since most master's degree recipients—about two-thirds—work in nonacademic settings, the utilization of the master's degree in the workplace is very different from the utilization of the doctorate.

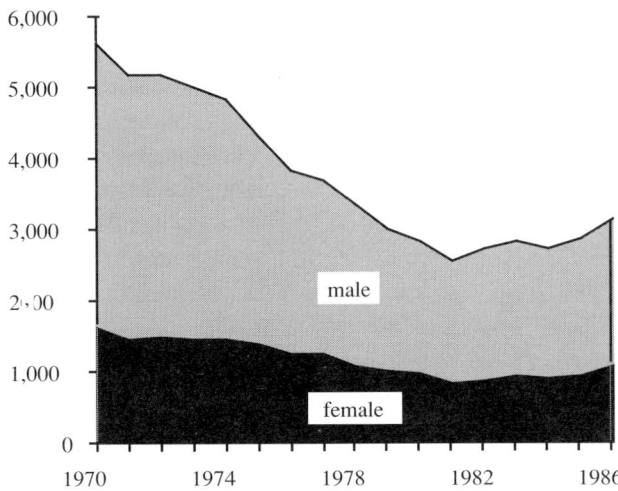

FIGURE 4.14 Master's degrees awarded, mathematical sciences. SOURCE: National Center for Education Statistics (NCES, 1988a).

Most institutions that offer doctoral degree programs in mathematics or statistics also offer programs for master's degrees. In mathematics, in addition to the 155 institutions offering doctoral degrees, another 273 institutions grant a master's degree as the highest degree (see Box 3.2). In statistics, there are approximately 217 master's degree programs at 172 institutions.

In 1986 three-fourths of the 2,700 mathematical sciences graduate students in master's-granting (not doctorate-granting) institutions were enrolled part-time. Master's-granting institutions included a higher proportion of women and U.S. citizen graduate students, than did doctorate-granting institutions. However, the proportion of non-U.S. students enrolled full-time in master's-granting institutions increased from 13% in 1977 to 34% in 1986, or by about 700 students. The gender composition of the master's degree recipients changed from 20% women in the mid-1960s to 35% women in 1986. The major increase in the representation of women occurred during the period from 1965 to 1975; since then the number and proportion of degrees awarded to women have been relatively steady at about 1,000 and 35%, respectively (see Appendix Table A4.1).

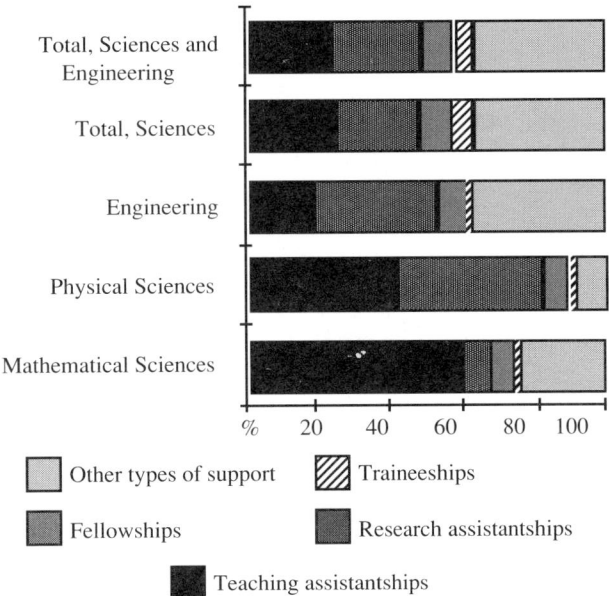

FIGURE 4.13 Types of major support for graduate students in doctorate-granting institutions, 1986. (See Appendix Table A4.8.) SOURCE: National Science Foundation (NSF, 1988a).

Majors in Mathematics and Statistics

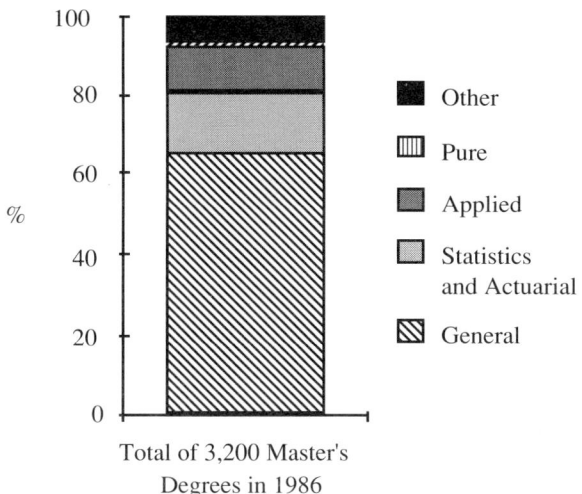

FIGURE 4.15 Master's degrees in mathematical sciences, distribution by subfield. SOURCE: National Center for Education Statistics (NCES, 1988a).

By subfield, two-thirds (65%) of the 3,200 mathematical sciences master's degrees awarded in 1986 were in general mathematics, about 16% were in statistics or actuarial science, and 12% were in applied mathematics (Figure 4.15). This distribution has not changed much in recent years except for increases in applied mathematics since the early 1980s.

The ethnic and racial composition of U.S. citizens receiving master's degrees in 1985 was almost identical to the composition of graduate students enrolled full-time in doctorate-granting institutions in 1986. Although non-Asian minorities were more highly represented in part-time enrollments and in master's-granting institutions, their shares of degrees awarded were similar to their fractions of full-time graduate students in doctorate-granting institutions. Non-U.S. students received one-quarter of all master's degrees awarded in 1985 (Table 4.8); just four years earlier in 1981, their share of the master's degrees was 18%.

Mathematical scientists, like engineers, attain doctoral degrees after master's degrees at a relatively low rate compared to that for other scientists (see Chapter 2). But a significant proportion of master's degree recipients in the mathematical sciences do continue on to study for a doctorate. Based only on the numbers of degrees awarded at the three levels, about one in five bachelor's recipients continue on for a master's degree, and also one in five master's degree recipients continue on for a doctoral degree. Less than one in ten women continue on for a doctorate from a master's degree, but nearly one in four men do (Table 4.9). Among the already-reduced attainment rates for mathematical science students compared to students of the other sciences, the graduate degree attainment rate for women is exceptionally low. The attrition of women along the path from the bachelor's to the doctoral degree is significantly higher in the mathematical sciences than in other science fields (NRC, 1983).

Demand for mathematical scientists at the master's degree level has increased since 1976. According to a

TABLE 4.8 1985 master's degrees awarded in mathematical sciences programs

	U.S. Citizens							
	White	Black	Hispanic	Asian	Indian	Total U.S.	Foreign	Total
Total	1,873	53	49	164	7	2,146	685	2831
Men	1,170	34	28	108	4	1,344	499	1843
Women	703	19	21	56	3	802	186	988
Percent								
By race in the U.S.	87.3	2.5	2.3	7.6	0.3	100		
By citizenship						76	24	100

SOURCE: NCES, unpublished data.

A Challenge of Numbers

TABLE 4.9 Attainment rates of master's and doctoral degrees

	Percent Master's to Bachelor's (2-year lag)	Percent Doctorate to Master's (5-year lag)	Percent Doctorate to Bachelor's (7-year lag)
All natural sciences and engineering	22	21	5
Mathematical sciences	21	19	4
Men	24	24	5
Women	17	9	1.5

SOURCE: Appendix Table A4.11 and National Science Foundation (NSF, 1987b).

survey by the NSF, recent mathematics master's degree recipients had a very low general unemployment rate (1.5%) and a high employment rate in science and engineering occupations (90%) (NSF, 1987a). This compared with a general unemployment rate of 2.1% for all recent science and engineering master's graduates and an employment rate of 84% in science and engineering. Median annual salaries for 1984 and 1985 mathematics master's graduates showed men receiving $4,000 more than women in 1986, and industries on the average paid graduates $7,000 more than did educational institutions. Close to half of master's graduates worked in the business and industry sector, two-fifths worked in educational institutions, and the remainder worked in government or other sectors.

Doctoral Degree Recipients

The number of doctorates awarded by U.S. universities in the mathematical sciences rose from about 200 in 1950 to a high of about 1,000 in 1973 and has fallen since to about 750 (Figure 4.16). In 1973 nearly four of five of these doctorates were earned by U.S. citizens. In 1988 fewer than half were. In the period from 1977 to 1987, more than one-third of the approximately 9,000 doctoral degrees awarded in the mathematical sciences went to non-U.S. citizens.

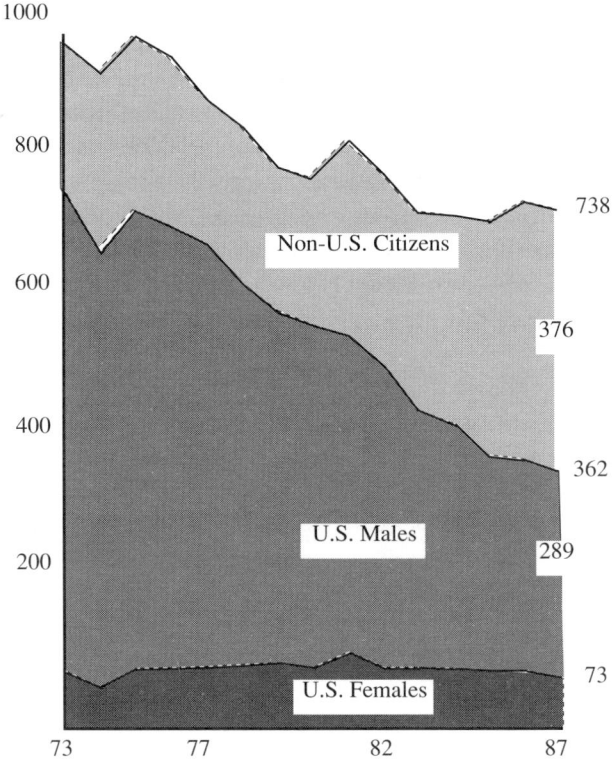

FIGURE 4.16 Ph.D. degrees in mathematics. (See Appendix Table A4.16.) SOURCE: American Mathematical Society (AMS, 1976 to 1987).

TABLE 4.10 Ratio of new doctorates in mathematics to new doctorates in selected other fields, 1970 to 1985

	Chemistry	Physics/Astronomy	Biology	Engineering
1970	0.55	0.74	0.36	0.36
1975	0.65	0.88	0.33	0.38
1980	0.48	0.76	0.20	0.30
1985	0.37	0.64	0.18	0.22

SOURCE: Edward A. Connors. Original data were taken from National Science Foundation (NSF, 1988c).

Certain of the trends in the number of mathematical sciences doctorates awarded are not unique; some of these same patterns have been observed in the total number of research doctorates awarded by U.S. universities. This total was about 10,000 in 1960, rose to a maximum of nearly 34,000 in 1973, and then fell slightly until it stabilized at about 31,000 in 1978. The percent of U.S. citizens among the recipients was nearly 85% in the early 1970s and had fallen to 72% in 1986. The number of doctorates awarded in the physical sciences, including the mathematical sciences, declined during the 1970s but increased during the 1980s. Statistics was an exception to the decline as degree production in that field was approximately stable, but in mathematics the decline was extreme. Mathematics is further exceptional in that it has been the slowest to stem the decline of the 1970s. The percent of U.S. citizens among new doctorates awarded in the physical sciences stood at 63% in 1986, about 12 points higher than in the subfield of mathematics.

Other comparisons point out that the decline in mathematical sciences doctorates since the mid-1970s is more exceptional. Edward A. Connors notes that each of the ratios of the number of new doctorates in mathematics to the numbers in chemistry, physics and astronomy, biology, and engineering has declined when computed for the years 1970, 1975, 1980, and 1985 (Connors, 1988). For example, in 1970 there were approximately half as many degrees in the mathematical sciences as in chemistry, but this had dropped to one-third as many in 1985 (Table 4.10).

A possible reason for this decline in ratios is the decision of those holding baccalaureates in mathematics to switch to doctoral study in other areas, and data are given to support the contention. The shifts increased during the late 1960s and through the 1970s but appear to have leveled off in the 1980s (Table 4.11).

The number of women U.S. citizens earning doctorates

TABLE 4.11 Mathematics majors going on to doctoral study in other areas of science and engineering, 1960 to 1985

	Mathematics majors earning doctorates in science/engineering[a]	Mathematics majors earning doctorates in mathematics	Ratio
1960	290	207	0.71
1965	639	484	0.76
1970	1362	924	0.68
1975	1310	883	0.67
1980	1139	609	0.53
1985	959	506	0.53

[a] Includes mathematics and the social sciences.
SOURCE: Edward A. Connors. Original data were taken from National Science Foundation (NSF, 1988c).

TABLE 4.12 Ethnic representations among new mathematical sciences doctorates, U.S. citizens, 1977 to 1986

	Number of new U.S. citizen doctorates in mathematical sciences	Percentage of new U. S. Citizen Doctorates	Approximate percentage of U.S. population
Asian-Americans	224	4.3	2
Blacks	80	1.5	12
Hispanics	34	0.6	7
Native Americans	26	0.5	0.5
Total, U.S. citizens	5,249	100%	

NOTE: For changes since 1974, see Appendix Table A4.12.
SOURCE: American Mathematical Society (AMS, 1976 to 1988).

each year in the mathematical sciences has been essentially constant since 1973, usually in the range 80 to 90. The percent of U.S. citizens receiving doctorates that is accounted for by this group has risen from 10% to 20% because the total number of U.S. citizens receiving mathematical sciences degrees each year has fallen from 774 to 362. Overall, among both U.S. and non-U.S. citizens, women constituted 10% of the new doctoral degree holders in 1973 and 17% in 1987. Among all doctorate recipients in 1986, women were better represented, having received 35% of the degrees; but in the physical sciences women were awarded only 16% of the new doctoral degrees and in engineering only 7%.

Blacks and Hispanics receive inordinately few of the doctorates awarded in the mathematical sciences. AMS Annual Survey results showed that during the ten-year period from 1977 to 1986, these two groups combined received about 2% of the doctorates awarded to U.S. citizens (Table 4.12).

Analogous data (NRC, 1987) for all research doctorates awarded by U.S. universities in the ten-year period from 1977 to 1986 are given in Table 4.13. Additionally, since the NRC report includes information on the visa status of non-U.S. citizens, a second comparison is allowed. Including those non-U.S. citizens with permanent visa status changes the representations very little except when Asian-Americans are included. Then the change is dramatic. Asian-Americans constituted 46% of the non-U.S. citizens with permanent visas who received a doctorate from a U.S. university between 1977 and 1986.

TABLE 4.13 Ethnic representation among all new research doctorates, U.S. citizens and permanent residents, 1977 to 1986

	Number of doctorates to U. S. citizens	Percentage of U.S. citizen doctorates	Number of doctorates to U. S. citizens or permanent residents	Percentage of U.S. citizen or permanent resident doctorates
Asian-Americans	4,579	1.9	10,404	4.2
Blacks	9,903	4.2	10,821	4.3
Hispanics	4,970	2.1	5,697	2.3
Native Americans	790	0.3	792	0.3

SOURCE: National Research Council (NRC, 1987).

Majors in Mathematics and Statistics

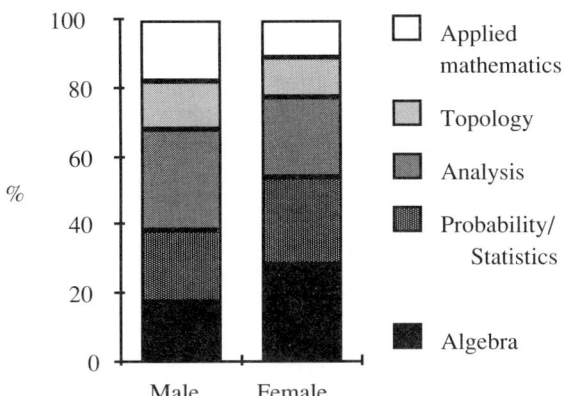

FIGURE 4.17 Doctoral degrees in mathematical sciences, distribution by subfield and sex. SOURCE: National Science Foundation (NSF, 1983); see Appendix Table A4.15.

The five major subfields in the mathematical sciences are algebra, probability and statistics, analysis, topology, and applied mathematics. These five subfields accounted for 65% of all doctorates awarded to both men and women in the period from 1960 to 1982. The most popular subfields in the mathematical sciences, as measured by the number of degrees awarded, have not been the same for men and women (Figure 4.17). In the two decades beginning in 1960, the three most popular subfields for women were algebra, probability and statistics, and analysis. For men the most common choices were analysis, probability and statistics, and applied mathematics. Women were more than twice as likely as men to opt for either algebra, probability and statistics, or analysis rather than for applied mathematics.

A broad interpretation of the mathematical sciences includes programs in other departments, such as operations research, computer science, statistics, and mathematics education. For example, operations research might be taught in departments of mathematics, engineering, or business. In the last decade, the total number of doctoral degrees awarded annually in the broadly interpreted area of mathematical sciences has been relatively constant at about 1,400 (Figure 4.18). The trend has involved student shifts away from mathematics and education and toward computer science, with the numbers choosing statistics and operations research remaining steady.

The median time from entry into graduate school to receipt of the doctoral degree was 7.3 years for the 1986 mathematical sciences doctoral degree recipients and has been essentially the same since 1958. The median for all fields was 10.4 years in 1986. Generally, this median is in the 7- to 9-year range for the sciences and engineering and in the 9- to 13-year range for the social sciences, the humanities, and education. The median registered time, which is the time spent from entry until completion of enrollment in graduate school, was six years for those who received mathematics doctorates in 1986.

Three-quarters of the 1986 mathematics doctoral degree recipients had their bachelor's degrees in mathematics, and the same fraction (73%) had master's degrees. In all the physical sciences, 73% of those with doctorates had bachelor's degrees in the same field, but only 52% had master's degrees. For all fields these percents were 55% and 79%, respectively.

About 200 postdoctoral appointments in mathematical sciences were made at doctorate-granting institutions in 1986. In recent years the number of appointments has ranged from a low of 110 in 1981 to a high of 225 in 1985 (NSF, 1988a). An NRC survey showed that the number of doctoral degree holders with postdoctoral study plans increased from about 10% to 23% of the total in the past ten

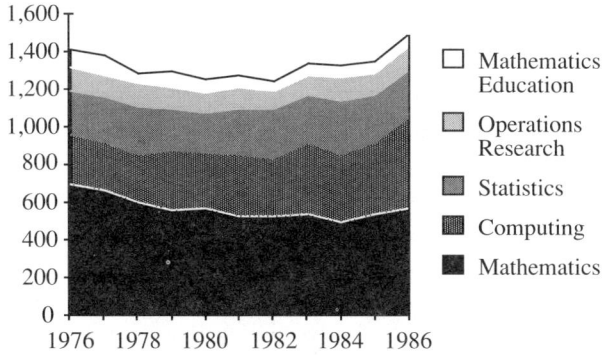

FIGURE 4.18 Number of doctorate recipients in broadly interpreted mathematical sciences. (See Appendix Table A4.14.) SOURCES: National Research Council (NRC, 1987) and American Mathematical Society (AMS, 1976 to 1987).

years (NRC, 1987). This is still far below the 40% in all the physical sciences who planned postdoctoral study. The NRC survey asked new doctoral degree recipients to check one of five reasons as the most important reason for either taking or deciding against a postdoctoral appointment. In this survey, two dominant reasons were given for deciding against postdoctoral study in mathematics: "No postdoctoral available" was given by 38%, and "attractive employment" was given by 42%. The other three choices were "little or no benefit," "inadequate stipend," and "other." The "no postdoctoral available" response was given most often in mathematics (38%) and the next most often in the humanities (31%); the response for all fields was 20%. The dominant reason given in favor of postdoctorals by those with mathematics doctorates was gaining additional experience, the reason given by 69% of those planning postdoctoral study. This reason too, led in all fields, with the overall percentage at 56%.

Excluding those who have postdoctoral positions, about three-quarters of doctorate mathematical scientists are currently employed in academe. The fractions taking nonacademic employment are higher for applied mathematics and for statistics and lower for mathematics. There have not been major changes in the employment patterns of doctoral degree holders over the years. Approximately 20% find employment in business and industry and the rest are employed in government. An increased demand in academia for mathematical scientists is likely because of the current low supply and an increase in retirements.

Patterns and Prospects

At all degree levels, several patterns are apparent. The numbers of degrees awarded annually have decreased significantly since peaking in the early 1970s but have shown some increases recently. The attrition rates in degree programs are high. Relatively few women, blacks, and Hispanics receive degrees, especially graduate degrees, and non-U.S. citizens are close to achieving a majority among graduate students. There are increased demands for mathematical sciences degrees in the workplace, and these increases are projected to continue.

 Mathematical Scientists in the Workplace

- **Three-fourths of those with mathematical sciences degrees are not classified as working as mathematical scientists.**

- **The number of persons identified as working as mathematical scientists almost tripled in the past decade.**

- **One-fourth of those with a bachelor's degree, one-third of those with a master's, and three-fourths of those with a doctorate begin work in educational institutions.**

- **White males currently dominate employment, and there is an increasing dependence on foreign nationals.**

- **The current supply of faculty depends heavily on persons from outside the United States, and the expected future supply will be insufficient to replace retirees.**

- **Shortages of qualified school mathematics teachers have developed, and various projections on needed replacements are alarming.**

Introduction

Standard U.S. labor market data do not tell much about where persons with mathematics or statistics degrees work or what they do. Over the past 40 years, approximately 525,000 persons have been awarded baccalaureate degrees in mathematics or statistics, 110,000 have received master's degrees, and 24,000 have been granted doctorates. Current estimates place the number of mathematical scientists in the work force at between 100,000 and 150,000, about one-fourth of what would be expected based on the number of degrees conferred.

Three factors account for much of this discrepancy. The title of mathematician has not been generally regarded as a professional title but is gaining recognition as one; statistics has been regarded as a profession only in recent years, but workers with the title of statistician may not hold college degrees in mathematics or statistics. Persons with degrees in mathematics work under various job titles; statistician, computer specialist, engineer, analyst, and actuary are common. And many such persons are secondary school teachers who are not usually classified as mathematical scientists.

The lack of information about the employment of people with mathematical sciences degrees reflects a separation between academic mathematical sciences and the nonacademic labor force. As the transition from high school mathematics to college mathematics is troublesome, so also is the transition from college to the workplace. The difficulty in making the latter transition is less understood, and building a better match between mathematical sciences education and the expectations and needs of the workplace is a major problem that is receiving increased attention.

Although general labor market data are of limited use in analyzing how mathematical sciences degrees are utilized, some trends are clear. The number of identifiable mathematical scientists in the work force has increased dramatically in recent years and is expected to continue to do so. A larger proportion of these degree holders are working in science and engineering fields. And mathematical scientists depend much more heavily on academic employment than does any other group in science and engineering.

Approximately 50,000 mathematical scientists are employed by college and university mathematical sciences departments, but only half of these are full-time faculty. This relatively small work force plays an essential part in educating much of the college-educated work force, especially students in science and engineering. The employment market for this faculty is unusual because there is no pool of readily available and well-qualified reserve candidates, and supply and demand have little effect on employment conditions. Proper balancing of the supply and demand has been difficult; currently there are mild shortages of candidates for the mathematical sciences faculty, and serious shortages are forecast. Obviously the conditions of supply and demand could resolve shortages, but planning and commitment will be necessary to maintain and enhance the quality of mathematical sciences education.

There continue to be shortages of secondary school mathematics teachers, and many who teach have inadequate preparation. Projections of the number of replacements needed by the year 2000 are alarming, in light of current trends and employment conditions.

As is true for the study of mathematics in colleges and universities, relatively few women, blacks, and Hispanics work in mathematics-based occupations. Their numbers are extremely low on college and university faculties where doctoral degrees are held by most of the members. Among secondary school mathematics teachers, the representation of women has improved significantly in recent years to about one-half.

General Characteristics and Trends

National labor statistics show that mathematical scientists account for less than 3% of the nation's total science and engineering work force, and the science and engineering work force constitutes about 4% of the total labor force of 120 million. In the last decade the science and engineering work force increased at an annual rate that was more than triple the rate for the general labor force, 7% versus 2%. The number working as mathematical scientists almost tripled in the decade ending in 1986. This was the largest increase for any science and engineering field, with

TABLE 5.1 Estimates of the number of mathematical scientists by National Science Foundation (NSF), Bureau of Labor Statistics (BLS), and Conference Board for Mathematical Sciences (CBMS)

	NSF (1986)	BLS (1986)	CBMS[a] (1985)
Employed in science and engineering	103,000	76,600	
Educational institutions	52,800	29,000	40,000 (25,000 full time)
Business and industry	35,600	37,700[b]	
Federal government and other	10,700	9,900[b]	
Other[c]	27,100	38,000	
Total	131,000	114,600	

[a] Includes faculty only, not graduate assistants.
[b] Includes mathematicians (20,000), statisticians (18,000), and actuaries (9,400).
[c] Includes non-S/E employed under NSF and operation researchers and analysts under BLS.
SOURCES: National Science Board (NSB, 1987), Bureau of Labor Statistics (BLS, 1988a), and Conference Board of the Mathematical Sciences (CBMS, 1987).

the exception of computer science (see Appendix Table A5.1). This increase in demand is expected to continue: from 1986 to 2000 the increase in demand for scientists, engineers, and technicians is projected to be 36%, and for mathematical scientists, 29%. This compares with a projected 19% increase in demand for the general labor force.

Estimates of the number of people working as mathematical scientists in the United States today range from 114,000 to 131,000, while the count of mathematical scientists employed in U.S. educational institutions ranges from 29,000 to 53,000. Some counts include only faculty members in mathematical sciences departments, whereas others include workers and faculty members in other departments. Moreover, the definition of mathematical scientists can vary. For comparison, three mathematics organizations (AMS, MAA, and SIAM) have a combined membership of approximately 46,000 people, and the two statistical societies (ASA and IMS) have a combined membership of approximately 17,000 (see Box 3.1). Table 5.1 gives estimates from three sources of the number of people working as mathematical scientists.

None of the estimates given in Table 5.1 appears to count elementary and secondary school mathematics teachers among the mathematical scientists. Estimates of their numbers vary, but the range is approximately 125,000 certified secondary school teachers plus 20,000 certified elementary and middle school mathematics teachers (NSB, 1985) to 300,000 public and private school mathematics teachers. In any event, there appear to be at least as many secondary school mathematics teachers as there are total mathematical scientists in the reported counts of science and engineering personnel. Clearly these teachers are not included as mathematical scientists, although many have the equivalent of a bachelor's degree in mathematics and work 90% of their day in mathematics. On the other hand, many persons with bachelor's degrees in science or especially in engineering are included in the counts.

Participation in the labor force and other selected employment characteristics for mathematical scientists were similar to those for all scientists and engineers in 1986. Most scientists and engineers, including mathematical scientists, are in the labor force, and the majority are

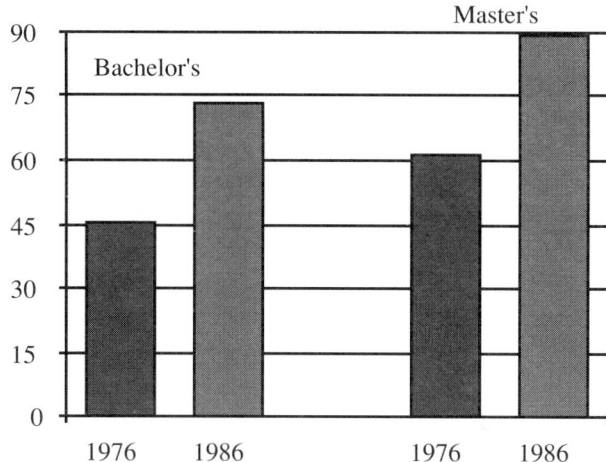

FIGURE 5.1 Percent of recent mathematics degree holders employed in a science or engineering job, 1976 and 1986. SOURCE: National Science Board (NSB, 1987).

working in science and engineering fields. The unemployment rates are low—between 1.1 and 2.4% (see Appendix Table A5.2).

Primarily a mental discipline, mathematical sciences is considered to be a field well suited to those with physical disabilities. However, only a small portion of mathematical scientists (1,600 of 131,000, or 1.2%) report a disability, compared with the 2% of all scientists and engineers who perceive themselves as disabled (NSF, 1988d). Disabled persons are distributed among the various fields at a rate similar to that for all scientists and engineers. Little information is available concerning working conditions for disabled people in the mathematical sciences, but the general impression is that there are fewer barriers in mathematics than in other fields. Advances in computer technology are expected to further increase opportunities and provide links for the disabled in all science and engineering fields (NRC, 1989).

The vast majority (85%) of science and engineering graduates find employment in a science and engineering field. The major employers include business and industry, educational institutions, and the federal government. One-half of all scientists work in business and industry, but only one-third of mathematical scientists do, whereas fewer than one in three scientists work in an academic setting, but half of mathematical scientists do (see Appendix Table A5.3). At the doctoral level, about three-quarters of mathematical scientists work in educational institutions; this fraction drops to about one-third at the master's level and to one-fourth at the bachelor's level.

Over the past ten years, the percentage of mathematics degree recipients employed in science and engineering fields has increased dramatically, from 46% to 74% for bachelor's degree holders and from 62% to 90% for master's degree holders (Figure 5.1); for all mathematical scientists, the percentage is 79%. The current employment pattern for mathematicians resembles that for computer scientists, physical scientists, and engineers, replacing the previous pattern, which resembled those in the life and social sciences, for which the percentages of bachelor's degree holders employed in science and engineering are somewhat lower (NSB, 1987).

Employment of Recent Graduates

Analyses of the employment patterns of recent recipients of mathematical sciences degrees reveal varied and improving opportunities that are related to the level of education achieved. Generally, the employer type shifts from business or industry to academe as the degree level rises from the baccalaureate to the doctorate. Exceptions are the recipients of doctorates in statistics or some applied mathematics fields; larger fractions of these people work in business or industry.

A survey of 1984 and 1985 bachelor's and master's degree recipients conducted by the NSF in 1986 showed that about half of all recent graduates at both degree levels worked in business or industry (NSF, 1987a). About one in four bachelor's degree recipients taught in educational institutions, presumably high schools. One-third of recent master's degree recipients taught in educational institutions, some in high schools, and others in colleges. According to a 1986 NRC survey of those new doctoral degree holders with employment plans (NRC, 1987), three of four planned academic work, one of five planned to work in business or industry, and fewer than one in ten planned to work in government (Table 5.2).

The majority of mathematicians with doctorates are members of college and university faculties, which are discussed in later sections of this chapter. This section's description of primary work activities and salaries pertains to master's and bachelor's degree recipients only. Although specific job titles and descriptions are not available, information on primary work activities and fields of employment sheds some light on what these workers do.

About 16% of the 1984 and 1985 bachelor's degree recipients surveyed in 1986 were enrolled full-time in graduate school, and another 13% were enrolled part-time. Full-time graduate students were excluded from the employment data. At both the master's and bachelor's degree levels, the major fields of employment were mathematics/statistics, computer science, and, to a lesser extent, engineering (Figure 5.2). A few mathematics graduates found employment in the fields of psychology and economics. As many bachelor's degree recipients found employment in computing science as in a mathematics or statistics field. About three of five master's degree recipients were working in a mathematics or statistics field; of the remaining

TABLE 5.2 Type of employer of mathematical scientists by degree level, 1986

	Bachelor's	Master's	Doctorates	All Math. Scientists	All Scientists
Educational institutions	26%	37%	73%	51%	29%
Business and industry	55%	48%	20%	34%	48%
Government and other	20%	15%	8%	15%	24%

SOURCES: National Science Foundation (NSF, 1987a) and National Research Council (NRC, 1987).

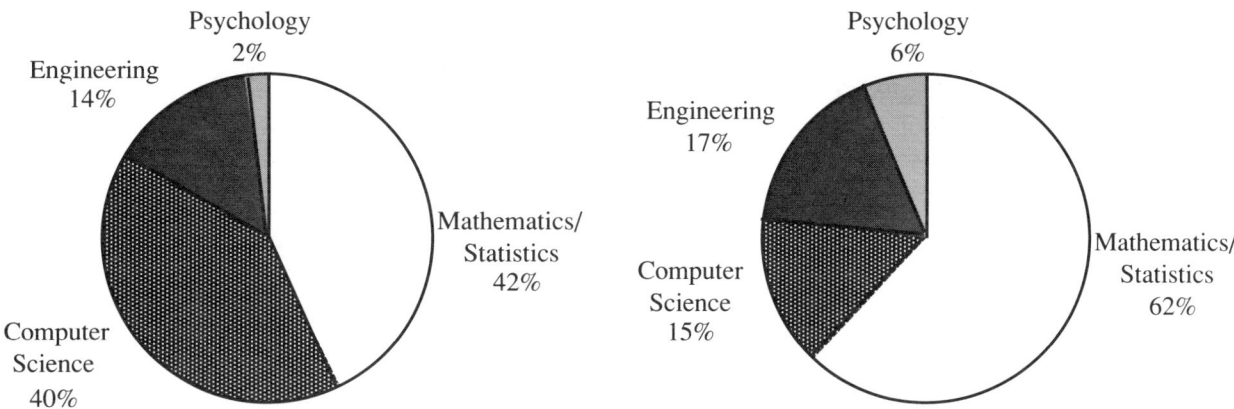

FIGURE 5.2 Field of employment for recent mathematics degree recipients, 1986. Left: Bachelor's degree holders. Right: Master's degree holders. SOURCE: National Science Foundation (NSF, 1987a).

two, one was working in an engineering field and the other in computer science.

Information on what nondoctoral mathematical scientists do in the workplace is very limited. Broad descriptive categories, such as primary work activities, field of employment, and type of employer, are available for the nondoctoral degree holder, but these broad categories offer little insight into what specific opportunities are available to these mathematics graduates. This type of information is needed to smooth the transition from college to the workplace for these graduates and to better meet the needs and expectations of business and industry.

A few colleges have information about what their mathematics and statistics graduates are doing, but local market conditions determine to a large extent what opportunities are available. Thus such information is not general, but rather only illustrative of opportunities. These opportunities include positions in educational, financial, governmental, religious, business, and industrial institutions. In educational institutions positions include mathematics teacher, coordinator for dropout prevention, guidance counselor, school principal, college instructor, and college professor. Some of the financial and business opportunities for mathematics graduates have included positions such as actuary, computer systems analyst, programmer, banker, bond specialist, insurance analyst, operations research analyst, financial analyst, financial accounting supervisor, pension consultant, employee education manager, and forecasting analyst. Other positions include lawyer, missionary, pastor, designer/draftsman, meteorologist, energy policy specialist, and marketing manager. Some of these positions require schooling beyond the bachelor's degree, but many do not. Contrary to many students' views of mathematics as too specialized for the workplace, students who have majored in mathematics are engaged in a wide variety of jobs with diverse work activities at different types of institutions.

The primary work activities of recent mathematical sciences degree holders have been separated into research and development, management and administration, teaching, production/inspection, reporting/statistical/computing activities, and other activities (Figure 5.3). As would be expected for graduates only two years out of college, fewer bachelor's than master's degree recipients were in management and administration positions in 1986, and more master's degree holders were teaching.

In 1986 the median annual salaries for recent mathematics degree recipients at both the master's and bachelor's degree levels were just slightly below the average for all science and engineering fields (Figure 5.4). The median annual salary for a bachelor's recipient was $24,100 and for a master's recipient, $31,500. For both groups the

A Challenge of Numbers

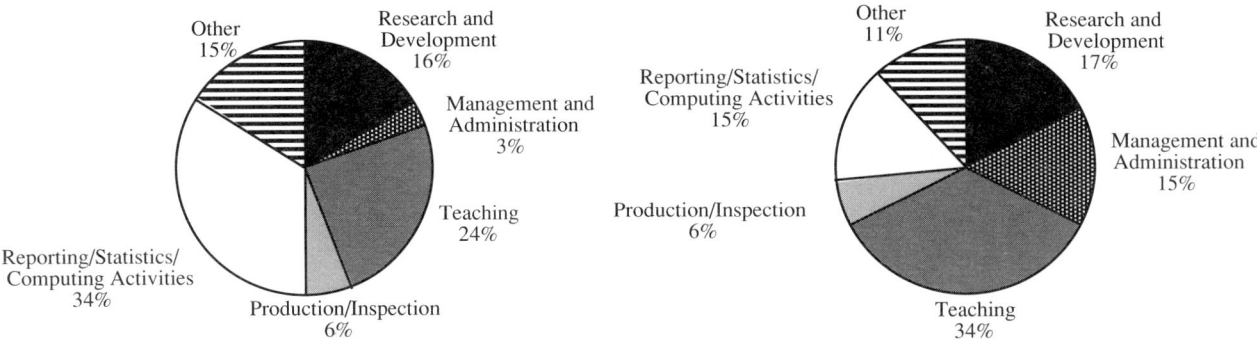

FIGURE 5.3 Primary work activities of recent mathematics degree recipients, 1986. Left: Bachelor's degree holders. Right: Master's degree holders. SOURCE: National Science Foundation (NSF, 1987a).

salaries were higher than those for physical, environmental, social, and life scientists but somewhat lower than those for engineers and computer scientists.

Secondary School Mathematics Faculty

The combined decrease in the number of both teachers and mathematicians graduating from colleges in the period from 1970 to 1985 (-39% for mathematics and -50% for education) has already resulted in a shortage of qualified mathematics teachers to staff the nation's schools. And more severe shortages are projected.

The decrease during the past decade in the number of college students planning to become teachers and an increase in the number of teachers approaching retirement predict shortages of teachers of all kinds. And losses from teachers leaving the profession in the middle stages of careers may further reduce the supply (OTA, 1988b, pp. 54-57). The general shortage of mathematical scientists together with the resulting demand across the work force adds to the prospects for too few secondary school mathematics teachers.

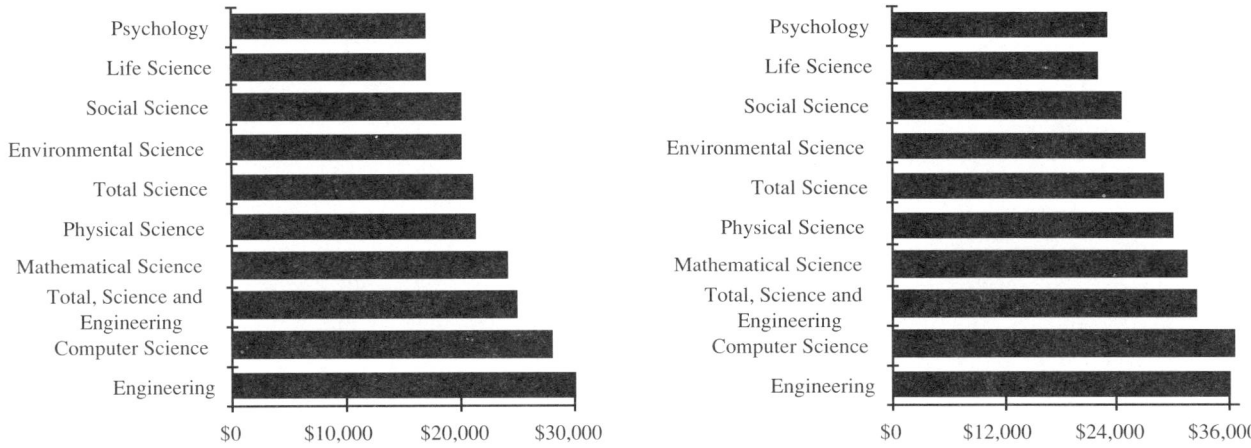

FIGURE 5.4 Median annual salaries of recent science and engineering graduates. Left: Bachelor's degree recipients. Right: Master's degree recipients. SOURCE: National Science Foundation (NSF, 1987a).

Approximately 11% of all secondary school teachers are mathematics teachers. The actual number of mathematics teachers in secondary schools is somewhat elusive but was estimated to be 126,300 in 1984 (NSB, 1985). Science teachers were estimated to number 115,600. Another estimate numbered the total teaching force of mathematics and science teachers at 200,000 in 1984 and estimated the number of new mathematics and science teachers needed by 1995 to be 300,000 (RAND, 1984). Thus, by the latter reckoning, the estimate of the new mathematics and science teachers needed in less than a decade exceeds the total current force. Other estimates of the current number of school mathematics teachers range as high as 300,000.

As the counts of mathematics secondary school teachers vary widely, so do estimates of shortages. However, most who have assessed the situation agree that future demand will be greater than current supplies, that the academic abilities of those attracted to education need to be elevated, that the most academically able are the most dissatisfied with teaching as a career, and that attrition is highest among the most able.

Demand for both secondary and elementary school teachers is projected to increase steadily from 1988 to the early 1990s. The supply of new teacher graduates is projected, at an intermediate projection level, to decrease slightly from 1988 to the early 1990s, leaving a deficit of 25,000 to 72,000 teachers each year (Figure 5.5).

The current shortages are not evenly distributed geographically or across disciplines, and the fields of mathematics and science have been particularly hard hit. In 1985 a low estimate of the shortages of mathematics teachers was 3,700 and of science teachers, 2,800 (NRC, 1985). If teachers currently assigned but not qualified were to be replaced at a modest rate of 5% of all teachers in the field, then the forecast of annual shortages would increase to 9,200 in mathematics and 8,000 in science. In 1987, according to opinions of teacher placement officers surveyed by the Association for School, College, and University Staffing, Inc., considerable shortages of mathematics teachers were reported in all regions of the country except the Northwest and the Rocky Mountain states (ASCUS,

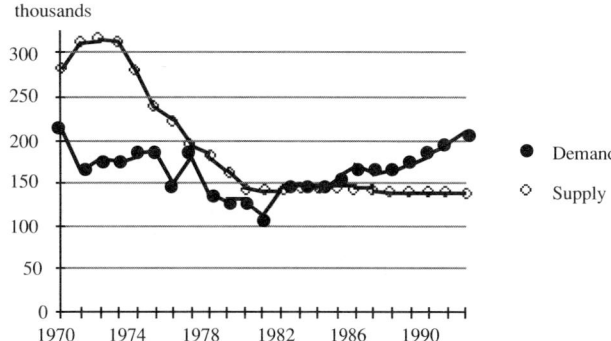

FIGURE 5.5 Supply and demand of new elementary and secondary school teachers, 1970 to 1992.
SOURCE: National Center for Education Statistics as reported in National Science Board (NSB, 1987).

1987). These shortages have been classified as considerable (having been assigned values of between 4.25 and 5.00 on a scale of 5.00) each year for the period from 1982 to 1987.

The *Report of the 1985-86 National Survey of Science and Mathematics Education* by Iris Weiss (RTI, 1987) gives some characteristics of the mathematics teaching force. The mathematics teaching force closest to college and university mathematics programs is the high school teaching force, which is the focus of this section of the report.

In the decade ending in 1986 the fraction of men on senior high school mathematics faculties dropped from two-thirds to only about one-half. Currently the vast majority (94%) of such faculty are white, 3% are black, 1% are Hispanic, and 1% are Asian. The average age is 40 and the average number of years of teaching experience is 14.2. A high school mathematics teacher is slightly more likely than not to have earned a degree beyond the bachelor's.

Approximately one-quarter of high school mathematics teachers do not have a degree in a mathematics or mathematics education field, but only 15% report teaching courses that they are not certified to teach. This compares with 16% of science teachers with a degree in a field other than science or science education, and 20% of science teachers who reported teaching courses that they are not certified to

A Challenge of Numbers

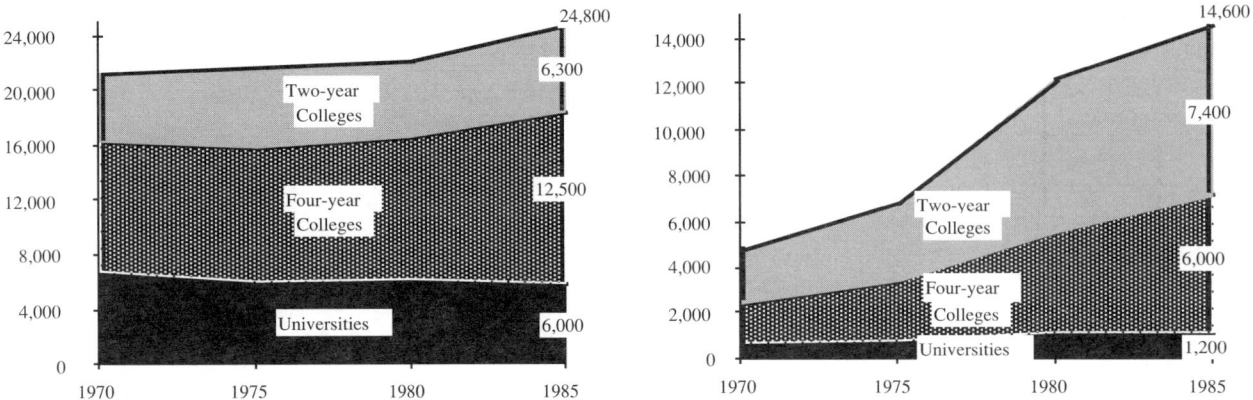

FIGURE 5.6 Left: Number of full-time mathematical sciences faculty members at colleges and universities. Right: Number of part-time mathematical sciences faculty members at colleges and universities. SOURCE: Conference Board of the Mathematical Sciences (CBMS, 1987).

teach. Most mathematics (84%) and science (89%) teachers are certified in their respective fields. Certification is determined locally, varies from district to district, and does not imply a uniform set of qualifications across the states.

The National Council of Teachers of Mathematics (NCTM) has developed guidelines for the preparation of mathematics teachers. The guidelines provide lists of competencies and recommend courses to develop these competencies for prospective teachers of mathematics. For instance, at the senior high school level the recommended courses for mathematics teachers include, among others, at least three courses of calculus, one of computer science, and two courses in methods of teaching mathematics. According to the Weiss survey, 36% of high school mathematics teachers have not completed three courses in calculus, 27% have not completed a course in computer science, and 46% have not completed two courses in methods of teaching.

Perceptions of the quality of mathematics high school teachers, based on self-evaluation and ratings by principals, reveal that most mathematics teachers enjoy teaching (95%) and agree that they are "master" mathematics teachers (63%). But their colleagues in science and in social science and history are more likely to be rated highly competent by the principals than are mathematics teachers, with percentages of 72% versus 67%. Of those mathematics teachers not rated highly competent by the principals, 30% were considered competent and 3% incompetent.

In the Weiss survey (RTI, 1987), the most frequently cited factors considered to be serious problems for mathematics teachers were student related. Almost one-fourth of senior high school mathematics teachers felt students' lack of interest in science, inadequate reading abilities, and absences were serious problems in their schools. Other less frequently cited serious problems were lack of materials, insufficient funds, large class sizes, and inadequate access to computers.

In addition to student-related problems, there are a host of other sources of discontent that were not included in the Weiss survey. *The Coming Crisis in Teaching* reports the results of a Rand study (RAND, 1984) that queried teachers about their views of the workplace. Between 40% and 50% of teachers who had degrees that reflected their area of teaching were dissatisfied because of a lack of administrative support, bureaucratic interference, a lack of autonomy, salaries, and other working conditions. Education majors also registered a certain amount of discontent with these same working conditions, but much less frequently (at percentages ranging from 5% to 25%) than did academic majors. Thus the most academically qualified teachers were also the most dissatisfied and, because of this, are more likely to leave teaching.

Many policymakers and educators point to low salaries as a major stumbling block to both improving the quality and increasing the number of mathematics teachers. Since salary schedules are generally the same for teachers regardless of the subject, salary levels are more critical in high-demand areas such as science and engineering. Differential salary levels are being considered to address teacher shortages but have not yet been widely implemented. In real terms, average annual public school salaries fell during the 1970s and by the mid-1980s were almost back to the 1970 level. The mean salary for teachers in 1986 was about $25,000, with wide variations among the states (OTA, 1988b, p. 57). The average starting salaries for public school teachers were $8,233 in 1975 and $16,500 in 1986. These compare to average starting salaries for college graduate mathematical scientists in private industry of $10,980 in 1975 and $23,976 in 1986 (BOC, 1988a). Public school salaries are generally for 9 or 10 months. If a 9-month salary is translated to a 12-month basis, the 1975 average salaries for teachers and employees in private industry were essentially the same. By 1986, however, even the 12-month equivalent of the teacher salary was about 10% less than the industry salary.

Both a high attrition rate, 9% in 1983, and a high retirement rate, estimated to be over 40% from 1983 to 1993, of mathematics and science teachers signal major replacement problems in the next few years (RAND, 1984). School districts may have to replace mathematics teachers with teachers from other fields. Furthermore, increases in the demand for secondary school mathematics teachers are likely because of increased high school graduation requirements in mathematics and fewer collegiate remedial programs in mathematics. If current patterns persist, then the prospects for a sufficient number of qualified replacements are dim.

Characteristics of College and University Faculties

The college and university mathematical sciences faculty, which accounts for about 5.5% of all faculty, numbers approximately 50,000, including about 8,000 graduate assistants and 15,000 part-time members. This small but critical work force has fundamental responsibilities for the education of many U.S. workers, bears the principal responsibility for maintenance and development of the disciplines of mathematics and statistics, and is charged with educating replacements and additions within its own ranks. These responsibilities have changed dramatically over the past 40 years, and there are major challenges projected by the year 2000.

Approximately three of every four mathematical sciences doctoral degree holders and about one of every four master's degree holders are on these faculties, most of whose members have their highest degrees in mathematics, mathematics education, or statistics. Among these faculty members are approximately 10,000 of the nation's estimated 11,000 research mathematical scientists. Approximately two-thirds of the 25,000 full-time faculty members are tenured.

Balancing the supply and demand for faculty members has been difficult because there have been no candidates in reserve—there are very few postdoctoral positions, and mathematics is a field people leave rather than move into. For many years, members of this faculty have come from other countries, and that practice has increased since the early 1970s as the number of U.S. citizens receiving mathematical sciences doctorates annually has dropped from over 900 to under 400.

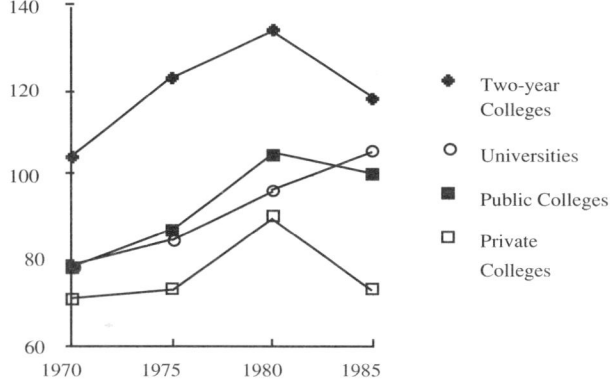

FIGURE 5.7 Mathematical and computer sciences enrollments per FTE of faculty. SOURCE: Conference Board of the Mathematical Sciences (CBMS, 1987).

TABLE 5.3 Professional activities of four-year college and university mathematical sciences faculty

	Mathematics			Statistics
	Universities	Public four-year colleges	Private four-year colleges	Universities
Classroom teaching performance	70 (3)	81 (2)	96 (4)	71 (6)
Published research	96 (0)	70 (10)	26 (39)	100 (0)
Service to department, college, or university	31 (5)	63 (5)	66 (0)	31 (11)
Talks at professional meetings	42 (5)	49 (11)	13 (28)	25 (11)
Activities in professional societies or public service	22 (8)	45 (4)	33 (9)	31 (6)
Supervision of graduate students	34 (7)	21 (32)	—	81 (0)
Undergraduate/graduate advising	9 (22)	24 (20)	39 (12)	21 (21)
Expository and/or popular articles	22 (13)	37 (14)	14 (40)	14 (19)
Textbook writing	9 (35)	17 (35)	11 (58)	12 (50)

NOTE: The first number in each cell is the percentage of departments responding that the activity was very important. The number in parentheses is the percent of responses that indicated the activity was of little or no importance. The possible responses were 0,1,2,3,4, or 5 with 0 meaning "no importance" and 5 meaning "very important", and the others indicating gradations between these. The percentages given above for "very important" represent the 4 and 5 responses while "little or no importance" represents the 0 and 1 responses. By subtracting the sum of these percentages from 100, one can get the percentage of 2 and 3 responses.
SOURCE: Conference Board of the Mathematical Sciences (CBMS, 1987).

The mathematical sciences faculty is more than 80% male and more than 80% white, with half of the others being Asian-Americans. The age distribution, skewed by the heavy hiring in the 1960s, predicts an increased retirement rate by 2000. Projected shortages of replacement candidates and general demographic trends and reform scenarios indicate even worse shortages.

What Faculty Members Do

Most colleges and universities expect mathematical sciences faculty members to perform in three areas: teaching, service, and research. The definitions of these three areas vary across institutions, and the boundaries are usually blurred and do not include increasing responsibilities for planning and reporting.

Teaching usually means having sole responsibility for conducting classes and evaluating student performance, and the numbers of classes and students vary. Most full-time faculty members in two-year or four-year colleges teach three to five separate classes each semester, constituting 12 or more contact hours per week. Most university faculty members teach one to three separate classes each semester, with the one-course load being rare and the lower loads occurring most frequently at institutions where the research expectations are higher. Frequently at larger institutions, classes are large, and a faculty member lectures 100 to 300 students and has an assistant for grading, conducting recitation or problem sessions, and helping students outside of class. When no such assistance is provided, class sizes are mostly in the 30 to 50 range at public four-year colleges and universities and in the 20 to 30 range at private institutions. Introductory classes are generally larger and advanced classes smaller. On aver-

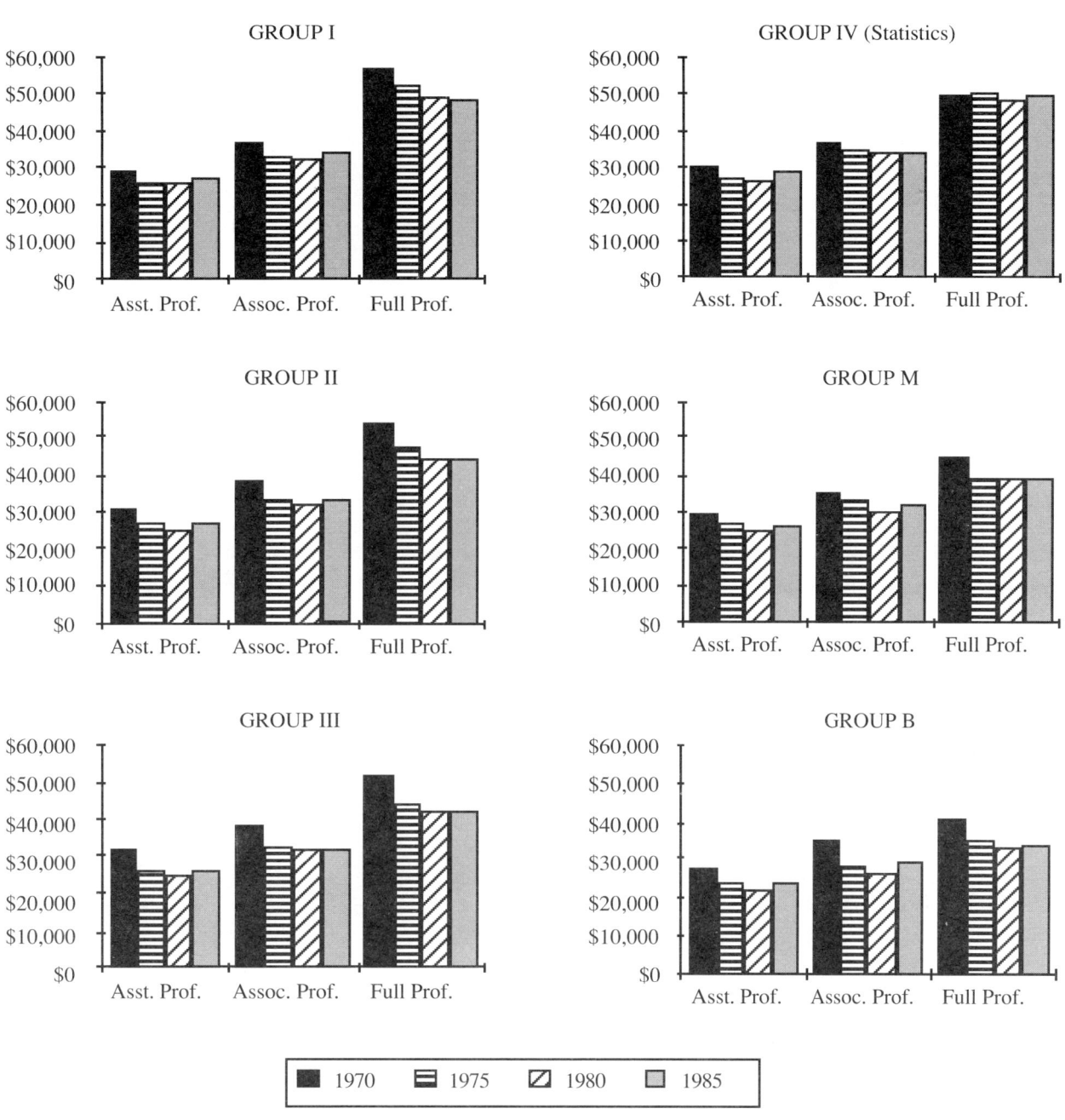

FIGURE 5.8 Mathematical sciences faculty salaries, 1970 to 1985 (in 1985 dollars). (See Box 3.2 for explanation of groups.)
SOURCE: American Mathematical Society (AMS, 1976 to 1988); see Appendix Table A5.13.

A Challenge of Numbers

age, faculty members teach about 120 students per semester in two-year colleges, 100 per semester in public four-year colleges, and 70 per semester in private four-year colleges. Other teaching duties include student advising, curriculum development, coordinating or supervising the teaching of others, and proposal writing and grant administration.

Service activities of faculty members include consulting, assisting public schools or community groups, institutional committee work, recruiting students and faculty members, and public relations.

The meaning of "research" varies from institution to institution. At research institutions, especially those with doctoral programs, the meaning of research in mathematics is usually clear. Research production usually means publishing new theorems on mathematics in refereed research journals or research monographs. In statistics, the meaning is usually broader, especially for applied statisticians. At other institutions, the definition of research production may be broader and may include textbook writing and expository writing, but there is no broader definition that is generally accepted by the academic mathematics community.

The CBMS surveys have asked department chairs to rate the importance of various professional activities in promotion or salary decisions. The most recent survey

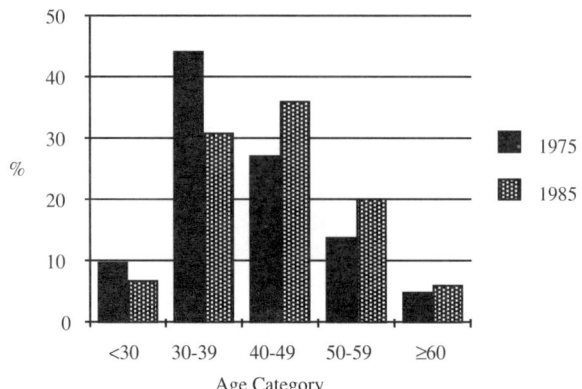

FIGURE 5.9 Age distribution of full-time mathematical sciences faculty in four-year colleges and universities. SOURCE: Conference Board of the Mathematical Sciences (CBMS, 1987); see Appendix Table A5.10.

results supported the emphasis on research at universities, with published research being most frequently cited as a very important activity (Table 5.3). At colleges, more importance was given to teaching activities.

For two-year colleges, the CBMS questionnaire did not ask the question summarized in Table 5.3 for four-year colleges and universities. The survey of two-year colleges asked for the percent of faculty engaging in certain profes-

TABLE 5.4 Professional activities of two-year college mathematical sciences faculty

	1975	1980	1985
Attending at least one professional meeting per year	47	59	70
Taking additional courses during year	21	22	31
Attending mini-courses or short courses	NA	NA	31
Giving talks at professional meetings	9	15	16
Regular reading of articles in professional journals	47	57	72
Writing of expository and/or popular articles	5	6	6
Writing research articles	NA	NA	3
Writing textbooks	15	10	4

NOTE: Numbers indicate percentage of faculty surveyed indicating participation in activity. NA means "not available."
SOURCE: Conference Board of the Mathematical Sciences (CBMS, 1987).

sional activities. The 1985-1986 CBMS Survey results showed that a maximum of 13% of the two-year mathematical sciences faculty wrote either textbooks or research, expository, or popular articles (Table 5.4).

An active program of scholarly activities outside of assigned teaching and service duties is generally accepted as necessary to keep faculty members intellectually alive and abreast of the developments in their discipline and in their profession. The losses from not having a program of "continuing education" or faculty development will be large if the discipline is changing and the curriculum is responsive to society's needs. The mathematical sciences are changing very rapidly, and a responsive curriculum should be a national imperative. When faculty members spend all of their working time conducting classes, especially low-level, routine courses, the losses in their effectiveness are accelerated through forgetting, boredom, and failing to keep up with new developments. A faculty member teaching for 35 years is likely to teach as many as 8,000 students, and so losses in teaching effectiveness can affect many students. In effect, intellectual capital (teaching potential and effectiveness) is being spent to conduct classes. Other professionals—for example, engineers in industry—consider continuing education essential to maintaining competence. This consideration is gaining broader acceptance and practice among academic mathematical scientists, and new patterns of professionalism are emerging.

Faculty Members by Duties and Credentials

In this section college and university mathematical sciences faculty are categorized by type of institution, academic credentials, full-time or part-time employment, subject area(s) taught, and research activity. The following subsets are useful in analyzing faculty characteristics and employment markets:

• *Doctorate full-time faculty members in four-year colleges and universities* (FT-D-FYCU) number approximately 16,000, with 6,500 in doctoral-degree-granting departments and 9,500 in master's-degree- and bachelor's-

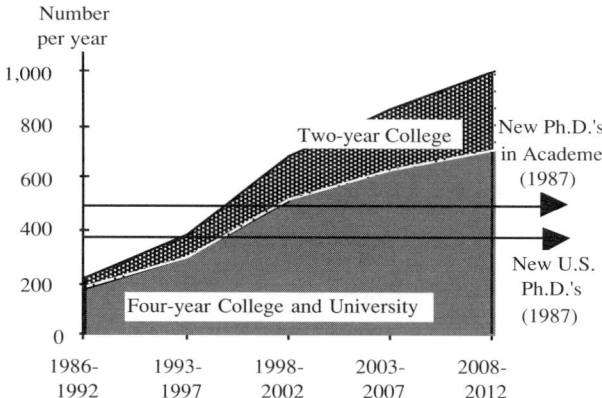

FIGURE 5.10 Estimated number of retirements of full-time college and university mathematical sciences faculty. SOURCE: Adapted from Conference Board of the Mathematical Sciences (CBMS, 1987) and American Mathematical Society (AMS, 1987).

degree-granting departments.

• *Research faculty members* publish traditional original research results regularly, and most have research as a designated part of their jobs. The size of this component is estimated at 10,000, constituting more than 90% of the nation's mathematical sciences researchers. Almost all are doctorate faculty at four-year colleges and universities, the bulk being at the universities.

• *Non-doctorate full-time faculty members in four-year colleges and universities* (FT-ND-FYCU) number approximately 4,500, with 500 in the doctoral-degree-granting departments and 4,000 in the others.

• *Part-time faculty members in four-year colleges and universities* (PT-FYCU) of whom there are approximately 7,000, not including graduate teaching assistants.

• *Full-time two-year college faculty members* (FT-TYC), a component that includes approximately 6,500 members; 13% have doctorates.

• *Part-time two-year college faculty members* (PT-TYC), who number approximately 7,500.

• *Graduate teaching assistants* (GTA), a group with approximately 8,000 members (all in doctoral and master's degree-granting institutions). Approximately 45% teach their own classes, another 40% conduct quiz or recitation sections, and the remainder perform other duties such as tutoring or grading papers.

Another basis for separating the college and university mathematical sciences faculty is by the fields in which they teach, notably mathematics, statistics, or computer science. (Although some faculty members work in other areas such as operations research, no finer classification will be made in this report.) The subsets of faculty given above separate into these disciplines approximately as given in Table 5.5.

The category "computer science" in Table 5.5 does not include faculty members in computer science departments, but only those in mathematical sciences departments. Also, although Table 5.5 includes many departments of statistics, there are units in statistics in other academic departments that are not included in this summary for mathematical sciences departments.

Comparing the number of full-time and part-time faculty members by category of institution over the period from 1970 to 1985 reveals little or no increase in full-time faculty members even though enrollments increased dramatically. In fact at universities the number of full-time mathematical sciences faculty in 1985 was 14% less than in 1970 (Figure 5.6; see Appendix Table A5.8). This represents a loss in both the mathematics faculty and the statistics faculty. Colleges and universities employed part-time faculty members to meet the demands of increased teaching loads, and this part-time sector of the faculty more than tripled from 1970 to 1985 (Figure 5.6). However, this trend has reversed; since the early 1980s the hiring of full-time faculty members has increased. An analysis of hiring for the period 1983 to 1988 is given in the section titled "Four-year College and University Doctorate Faculty."

The increase in part-time faculty members was not enough to keep pace with enrollments. From 1970 to 1985 enrollments per FTE of faculty member increased, especially during 1970 to 1980. At both two- and four-year colleges, enrollments per FTE appeared to level off, but at universities the number of enrollments per FTE steadily increased until 1985 (Figure 5.7). The data are combined for mathematical sciences and computer science in the CBMS reports, but the data for mathematics alone yield even higher numbers of enrollments per FTE.

These two trends, steady increases in part-time faculty and in enrollments per faculty member, have apparently been reversed in the 1980s. The heavy use of part-time faculty members was viewed by many as a serious problem, as reflected in the 1985-1986 CBMS Survey (CBMS, 1987). The need to use temporary and part-time faculty

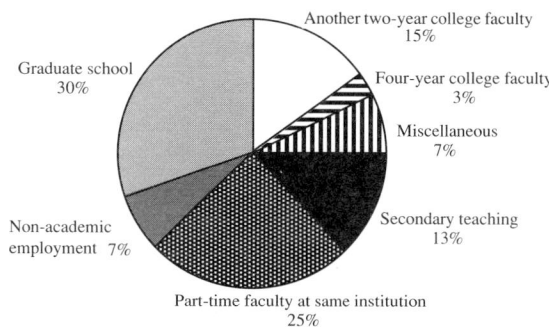

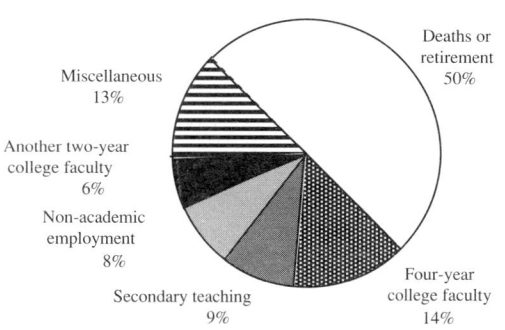

FIGURE 5.11 Top: Source of new hires of two-year college full-time faculty in mathematical sciences. Bottom: Destination of departing mathematical sciences two-year college full-time faculty. SOURCE: Conference Board of the Mathematical Sciences (CBMS, 1987).

TABLE 5.5 Numbers of mathematical sciences faculty members by teaching area and type of institution, 1987

	Mathematics	Statistics	Computer Science	Total
FT-D-FYCU	13,200	1,500	1,300	16,000
FT-ND-FYCU	3,750	50	700	4,500
PT-FYCU	3,950	50	3,000	7,000
FT-TYC	5,600	200	700	6,500
PT-TYC	6,450	250	800	7,500
GTA	6,250	1,750	0	8,000
Estimated FTE	27,600	2,300	4,000	33,900

NOTE: Each part-time faculty member is considered one-third of FTE and each GTA is considered one-quarter of FTE. Distribution of the faculty members is partially determined by the distribution of teaching responsibilities among the three disciplines. See "Faculty Members by Duties and Credentials," p. 65, for explanation of acronyms.
SOURCES: These data are based on Conference Board of the Mathematical Sciences (CBMS, 1987) and American Mathematical Society (AMS, 1976 to 1988).

members was classified as a very important problem by 35% of the responding university statistics departments, 42% of the university mathematics departments, 44% of the public four-year college mathematics departments, 42% of the private four-year college mathematics departments, and 61% of the two-year college mathematics departments.

The development of computer science was a contributing factor to faculty hiring practices in mathematical sciences departments in the period from 1970 to 1985. As computer science enrollments and the number of majors increased rapidly, positions—full-time and part-time—became available. Universities were much more likely to have a separate computer science department than were the four-year or two-year colleges. Thus, especially from 1970 to 1982, many faculty members were hired to teach computer science, not mathematics or statistics, which would partially account for the decline in the number of full-time faculty in universities while four-year full-time faculties were growing, albeit slowly. Although verifying data are not available, it is likely that the number of full-time faculty members teaching mathematics declined significantly in the period from 1970 to 1982, with the growth of computer science more than absorbing the new faculty hiring.

The Research Faculty

Most of the active mathematics and statistics researchers in the United States are in the doctorate-granting programs in universities. The 1984 David Report estimated the mathematical sciences research community at 10,000, with 9,000 of these being faculty members in educational institutions and having research as their primary or secondary activity (NRC, 1984). Of the 9,000 researchers in academia, the David Report estimated that 5,500 published regularly, 4,000 frequently, and 2,600 on a highly productive schedule. Extrapolating those estimates to a larger faculty yields approximately 11,000 active researchers in 1987.

A 1986 NSF survey (NSF, 1986a) of top research institutions, which yielded responses from 105 mathematical sciences departments, showed an average per department of 40 full-time faculty, up 7% from an average of 37 in a similar 1974 survey and up from an average of 36 in 1980, again reflecting the increased hiring in universities after

A Challenge of Numbers

TABLE 5.6 Age distribution of mathematical sciences faculty members in 105 research universities, 1980 and 1986

	< 30	30-39	40-49	50-54	55-59	60-64	≥ 65
1980	7.9%	37.0%	31.2%	10.9%	7.5%	3.8%	1.7%
1986	6.7%	25.8%	36.4%	12.6%	9.9%	6.5%	2.1%

SOURCE: National Science Foundation (NSF, 1986a).

1980. The percent of faculty with a recent doctorate (received in the previous seven years) in the responding departments dropped from 35% in 1974 to 22% in 1980 and to 21% in 1986. At research institutions, some 5.5% of the doctorate faculty in the mathematical sciences were aged 60 or over in 1980; 8.6% were in that age category in 1986 (Table 5.6). Except for computer science, this was the lowest percent of faculty aged 60 and over for the 22 areas of science and engineering surveyed.

The 1986 NSF survey also showed that the number of women on the faculties of research universities was about half that on the faculties of all four-year colleges and universities, but their representation improved slightly from 1980 to 1986, from 7.1% to 8.5% of the total faculty (NSF, 1986a). In 1986 women constituted 5.1% of the senior doctoral degree holders, 11.5% of those with recent doctorates, and 48.9% of the nondoctoral degree holders. Blacks and Hispanics were also poorly represented on these faculties, but the representation of Asians was about the same overrepresentation, in terms of their percentage of the total U.S. population, as for all four-year institutions.

Mathematics topped all the sciences in the proportion (35%) of full-time assistant professors with foreign baccalaureates. Only mechanical and electrical engineering had larger proportions, and both of those were under 40%. Physics was the second highest of the sciences at about 25% (NSF, 1987c).

Faculty Salaries

Generally, wages increase as the demand for workers increases and as the supply decreases, but this is not so in the college and university mathematical sciences faculty. Normal economic analyses do not apply. In the period from 1970 to 1985, compensation decreased as the number of faculty candidates fell and the teaching responsibilities increased. In constant dollars, professors (assistant, associate, and full) were earning lower salaries in 1985 than in 1970. Across the board, almost without exception, from doctorate-granting to master's-granting to bachelor's-granting institutions, faculty members are not as well compensated today as they were 15 years ago (Figure 5.8).

Ages of Faculty Members

Heavy hiring in the 1960s of relatively young faculty members resulted in a faculty with more than half its members under age 40 in 1975. Decreased hiring since the 1970s has resulted in an older faculty (see Figure 5.9). Similarities exist in the age distribution for doctorate mathematical sciences faculty members (Table 5.6) and for all faculty at four-year colleges and universities, except

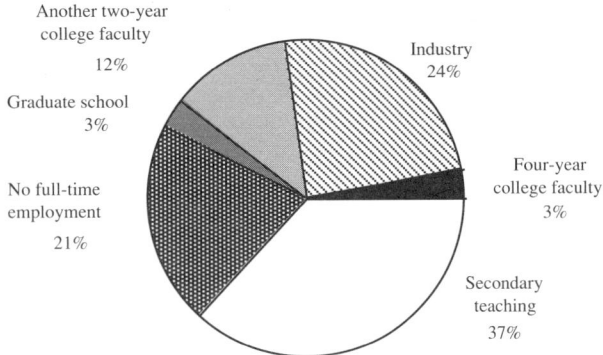

FIGURE 5.12 Source of two-year college part-time faculty in mathematical sciences. SOURCE: Conference Board of the Mathematical Sciences (CBMS, 1987).

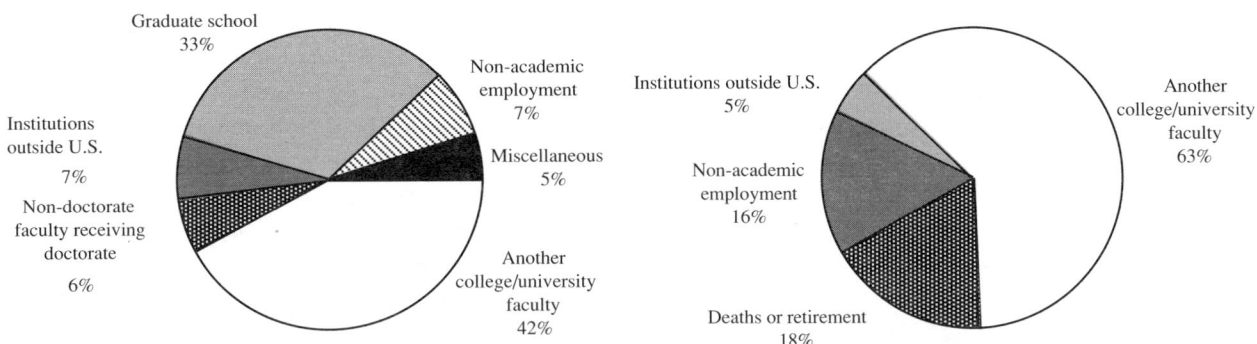

FIGURE 5.13 Left: Source of new hires of doctorate faculty in mathematical sciences, 1983 to 1988. Right: Destination of departing mathematical sciences doctoral faculty, 1983 to 1988.
SOURCE: Consolidated from AMS-MAA annual surveys (AMS, 1983 to 1988).

for a significantly larger fraction (18.5% in 1986) over the age of 55 at research institutions. In 1985 half the four-year college and university faculty members were 40 to 55 years old, as were more than half of the two-year college faculty members. The two-year college mathematical sciences faculty also had a very large fraction (24%) of its members in the 40 to 44 age bracket.

Estimates of retirements based on faculty age distributions indicate serious shortages of replacements by the year 2000 if present trends continue (see Appendix Tables A5.10 and A5.11) Over the past several years, the full-time mathematical sciences faculty at four-year colleges and universities has been retiring at rates of nearly 1% per year. The average number of retirements or deaths among the doctorate faculty for the five years from 1982 to 1987 was 142 per year, or about 1% per year. It is likely that these estimates for the period 1986 to 1992 are low because the data already available for 1986 and 1987 indicate a higher rate of attrition through death and retirement; in addition, early retirement options are increasing, and other sources of data give larger percentages for the population aged 60 and over. For example, NSF data give this percentage as 10%. For two-year college faculty, the estimate of retirements for the period 1986 to 1992, based on the ages of faculty in 1985, is much lower than the 217 retirements or deaths reported in the CBMS survey for the one year 1984-1985 (CBMS, 1987). Early retirements from this faculty are apparently popular, as there was

TABLE 5.7 Full-time mathematical sciences faculty by ethnic origin and sex, 1985

	Asian	Black	Hispanic	Native American	Men	Women
All four-year colleges	7.1%	3.5%	3.4%	0.1%	85%	15%
All two-year colleges	3%	4%	4%	1%	69%	31%
Statistics					90%	10%
University mathematical sciences					89%	11%
Public four-year mathematical sciences					81%	19%
Private four-year mathematical sciences					85%	15%

SOURCE: Conference Board of the Mathematical Sciences (CBMS, 1987).

considerable attrition in the group aged 50 to 60 between 1980 and 1985. The current level of production of U.S. Ph.D.s falls far short of the expected needed replacements of retiring faculty members in coming years (Figure 5.10).

Women and Minorities on the Faculty

Table 5.7 shows the fractions of women, blacks, and Hispanics on the college and university full-time mathematical sciences faculty (CBMS, 1987). The representations of women on the faculties at four-year colleges (15%) and at two-year colleges (31%) are very near their current representations in the recipients of new doctorates and new master's degrees. Among the various categories of institutions, universities have smaller fractions of each of women, blacks, and Hispanics on their mathematical sciences faculties.

Two-Year College Faculty Mobility

The 1985-1986 CBMS Survey (CBMS,1987) reported the results of a 1979 survey (McKelvey, Albers, Liebeskind, and Loftsgaarden) showing that more than 60% of all mathematics faculty in two-year colleges had previously taught in secondary schools, whereas the survey for 1985 showed that most new full-time hires came from either graduate school or from the part-time faculty at the same institution (Figure 5.11). The bulk (nearly 50%) of the outflow was due to deaths and retirements.

To explain the 1979 survey's finding that 60% of the faculty had previous experience in secondary teaching, one needs to look at the source of part-time faculty, a major feeder of the full-time faculty. For 1985 the major sources of 7,500 part-time two-year colleges mathematical sciences faculty were secondary school teachers and industry employees (Figure 5.12).

Four-year College and University Doctorate Faculty

The number of doctorate faculty in the mathematical sciences in four-year colleges and universities was approximately 16,000 in 1987 (CBMS, 1987). This faculty expanded from approximately 13,000 in 1975 and from 14,000 in 1980, reflecting a much faster rate of increase during the 1980s when institutions finally began to address the additional faculty needs brought on by increased enrollments. Various counts over the past years have included computer science faculty in computer science departments and in mathematical sciences departments. The count of 16,000 in 1987 did not include faculty from computer science departments, but it did include some faculty members (approximately 1,300) who taught computer science in mathematical sciences departments.

The number of new hires in this faculty has been in the range of 1,200 to 1,300 in recent years, with 500 to 600 of these being persons who have switched from one institution to another within this same faculty. Taking out this internal movement, the principal source of new hires into this faculty has been graduate school, and the principal reasons for leaving have been death and retirement and nonacademic employment (Figure 5.13).

The net result of the inflows and outflows to and from the mathematical sciences doctorate faculty since 1982 has been an average increase of about 400 members per year. The average net flow into this faculty can be organized into six categories, but five of the categories sum to zero, leaving the net increase as essentially the number hired from graduate schools (Table 5.8).

Four-Year College and University Nondoctorate Faculty

In 1987 there were approximately 4,500 nondoctorate full-time faculty members in four-year colleges and universities. This is essentially the same number as reported five years previously in 1982. Almost all of these faculty members were in the institutions granting master's and bachelor's degrees, with only about 400 reported as being in doctorate-granting departments.

In recent years the number of hires in this faculty has been 500 to 600 (AMS, 1987). About 20% of these have been persons switching institutions. The principal source for additions to this faculty has been graduate school, which has provided 60% of the new hires in the past five years. The other 40% have come from various sources, including the part-time faculty and the two-year college and high school faculties. The principal reasons persons left this faculty were to return to graduate school for doctoral work (19%), to move to the doctorate faculty after earning a degree and not moving (18%), and because of death and retirement (23%). There has been a small net outflow (a 1982 to 1987 yearly average of 19) from this faculty to nonacademic employment and a small net inflow (a 1982 to 1987 yearly average of 8) from institutions outside the United States.

Summary

Although about three-fourths of the people with college degrees in the mathematical sciences are not identified as mathematical scientists in the labor force, mathematical scientists are becoming more visible in the workplace. The number of workers so identified has tripled in the past decade, and the fraction of those working in science and engineering fields has increased dramatically. Still, although nonacademic employment is increasing, three-fourths of those with doctorates in the mathematical sciences begin work in an educational institution. These fractions are lower for applied mathematicians and statisticians.

Although white males dominate employment as mathematical scientists, particularly on the college and university faculty, there is an increasing dependence on foreign nationals. These situations leave the supply vulnerable to predictable shifts in demographics and unpredictable shifts in U.S. foreign relations.

TABLE 5.8 Estimate of average annual net flow into doctoral faculty at four-year colleges and universities, 1982 to 1987

	Net Flow
Graduate schools	400
Non-doctorate faculty	70
Non-academic employment	-50
Non-U.S. employment	50
Miscellaneous/unknown	100
Deaths and retirements	-150
Total	420

SOURCE: Consolidated from American Mathematical Society (AMS, 1976 to 1988).

There are currently shortages of secondary school mathematics teachers, and these shortages are expected to worsen in the future. Moreover the quality of teacher education is of great concern. The collegiate mathematical sciences faculty is aging and is of uncertain vitality. If present trends continue, and as retirements increase over the next decade, qualified replacements for both the elementary and secondary school and collegiate faculties will be in short supply. If the system were to be changed with the intent of improving the quality of mathematics instruction and scholarship, the shortages would be dramatic. For example, shortages would increase if teaching loads were reduced, more research support became available for graduate fellowships and postdoctoral salaries, or full-time faculty assumed more responsibility for teaching.

6 Issues and Implications

The challenge of numbers in the foregoing chapters is clear: How can the nation's growing need for mathematically skilled workers be met in the face of shrinking populations from which these workers have been traditionally drawn? General options include increasing the proportion of such workers from both traditional and nontraditional sources and increasing the utilization and the effectiveness of available workers. Specific actions that would lead to improvements are more difficult to identify, and formulations of these will be left to the final report of the MS 2000 Committee. However, some segments of the general challenge are formulated below.

The five previous chapters describe the people in the mathematical sciences from the perspective of college and university programs. These people—students, teachers, and other workers—are scattered throughout the educational system and the nonacademic workplace. The picture that emerges is strongly influenced by two general facts:

- The workplace is changing as jobs require higher-level skills and greater adaptability. Mathematics-based jobs are leading the way in increased demand.

- If present patterns persist, most socioeconomic and demographic trends indicate that fewer students will study mathematics and choose mathematics-based careers.

These trends point to an increased demand for and a shrinking supply of mathematical scientists and other mathematically educated workers. The nation must recognize this critical condition, and understand the major challenge it poses for U.S. education in general and for college and university mathematical sciences in particular. Educating workers for business and industry and teachers for all levels of education may require fundamental changes in a system already stressed by the events of the past three decades.

Several issues that require the nation's attention are apparent. These are raised by the following questions:

- *How can national needs for mathematically educated workers be met?* How can the expected shortage of mathematically trained workers be averted? How can available workers be better utilized? What incentives will attract more interest in mathematics-based occupations, especially among women, blacks, and Hispanics?

- *What changes are necessary to attract more students to the study of mathematics?* How can the mathematical sciences respond to the change in the traditional pool of U.S. college students? What and who will stimulate students to study the mathematical sciences? How can a more diverse group of students be attracted to mathematics, reducing the heavy dependence on white males? What are the consequences of heavy dependence on non-U.S. students in graduate programs? How can teaching become more effective and stimulating?

• *What can be done to improve the success rate of students during the transition from high school mathematics to college mathematics?* How can high school preparation and college expectations be better reconciled? What effects are remedial programs and overlaps between the content of high school courses and college courses having on student attrition?

• *What can colleges and universities do to meet the national need for school mathematics teachers?* What is the appropriate education for secondary school mathematics teachers and for elementary school teachers? How can the college and university faculty assist in implementing new standards for school mathematics? What program of continuing education for teachers will enhance school mathematics instruction?

• *What actions will spur renewal and revitalization of the mathematical sciences faculty?* What steps should be taken to ensure replacements for the aging collegiate faculty? What is appropriate preparation for collegiate teaching? What continuing program of scholarship for the non-research faculty is necessary to maintain the intellectual vitality of the profession? How can better compensation, incentives, and working conditions be achieved and maintained? What is necessary to maintain and enhance the research production of the faculty?

• *How can better monitoring of the mathematical sciences be implemented?* How can both professional organizations and government agencies cooperate in the collection and reporting of information? How can data be collected, organized, and disaggregated to provide a comprehensive view of the mathematical sciences community? How can mathematical scientists be identified in the nonacademic workplace?

• *How can colleges and universities prepare graduates who are more valuable and effective in the nonacademic workplace?* What changes would make mathematics graduates more valuable to business and industry? How can the full potential of the contributions of mathematical scientists be explored? What new educational programs could diversify the employment opportunities for mathematical scientists? Are there unrecognized opportunities for the Ph.D. in the mathematical sciences?

Although these issues center on the mathematical sciences enterprise in U.S. colleges and universities, they have implications for all of society. Monitoring and maintaining the health of this administratively decentralized and diverse enterprise transcend the normal roles and responsibilities of academic systems. These concerns and the importance of a continued healthy flow of mathematical talent are the reasons that the MS 2000 project and, in particular, this report were begun. The forthcoming descriptive reports on curriculum and resources and the prescriptive final report of the MS 2000 Committee will provide the nation with an agenda for revitalization of college and university mathematical sciences and with recommendations for continued monitoring and assessment.

Bibliography

AAAS (American Association for the Advancement of Science), 1984. *Equity and Excellence: Compatible Goals.* Office of Opportunities in Science, American Association for the Advancement of Science, Washington, D.C.

AAAS, 1988a. "AAAS Presidential Lecture: Voices from the Pipeline," Widnall, Sheila E., American Association for the Advancement of Science, *Science,* September 30, 1988, Volume 241, pp. 1740-1745.

AAAS, 1988b. "AMS Celebrates—and Worries," Cipra, Barry A., American Association for the Advancement of Science, *Science,* October 7, 1988, Volume 242, p. 34.

AAAS, 1989. "Wanted: 675,000 Future Scientists and Engineers," Holden, Constance, American Association for the Advancement of Science, *Science,* June 30, 1989, Volume 244, pp. 1536-1537.

ACE (American Council on Education), 1981. *Recruitment and Retention of Full-Time Engineering Faculty, Fall 1980.* Atelsek, Frank J., and Gomberg, Irene L., Higher Education Panel Report Number 52, American Council on Education, Washington, D.C.

ACE, 1984. *Student Quality in the Sciences and Engineering: Opinions of Senior Academic Officials.* Atelsek, Frank J., Higher Education Panel Report Number 58, American Council on Education, Washington, D.C.

ACE, 1985a. *Access to Higher Education: The Experience of Blacks, Hispanics, and Low Socio-Economic Status Whites.* Lee, Valerie, American Council on Education, Washington, D.C.

ACE, 1985b. *Conditions Affecting College and University Financial Strength.* Andersen, Charles J., Higher Education Panel Report Number 63, American Council on Education, Washington, D.C.

ACE, 1985c. *General Education Requirements in the Humanities.* Suniewick, Nancy, and El-Khawas, Elaine, Higher Education Panel Report Number 66, American Council on Education, Washington, D.C.

ACE, 1985d. *Recent Changes in Teacher Education Programs.* Holmstrom, Engin Inel, Higher Education Panel Report Number 67, American Council on Education, Washington, D.C.

ACT (American College Testing), 1989. *State and National Trend Data for Students Who Take the ACT Assessment.* American College Testing Program, unpublished data.

Albers, Donald J., Rodi, Stephen B., and Watkins, Ann E. (editors), 1985. *New Directions in Two-Year College Mathematics.* Proceedings of the Sloan Foundation Conference on Two-Year College Mathematics, Springer-Verlag, New York.

Albers, Donald J., Alexanderson, G.L., and Reid, Constance, 1987. *International Mathematical Congresses: An Illustrated History 1893-1986.* Springer-Verlag, New York, pp. 46-53.

Bibliography

AMS (American Mathematical Society), 1976 to 1988. Annual AMS Survey Reports 1976 to 1986, *Notices of the American Mathematical Society*. Annual AMS-MAA Survey Reports 1987 to 1988, *Notices of the American Mathematical Society*. Various Authors.

ASA (American Statistical Association), 1987. "U.S. and Canadian Schools Offering Degrees in Statistics," American Statistical Association, *Amstat News*, November 1987, pp. 71-83.

ASCUS (Association for School, College and University Staffing), 1987. *Teacher Supply/Demand 1987.* Atkin, James N., Career Planning and Placement Centers, Association for School, College and University Staffing, Inc., Addison, Illinois.

AWM (Association of Women in Mathematics), 1988. "Testimony to the Task Force on Women, Minorities and the Handicapped in Science and Technology," Ruskai, Mary Beth, Association of Women in Mathematics, *Association of Women in Mathematics Newsletter*, Volume 18, Number 4, July-August 1988, pp. 8-9.

Bacas, Harry, 1988. "Desperately Seeking Workers," *Nation's Business*, February, pp. 16-23.

Barzun, Jacques, 1988. "Multiple Choice Flunks Out," *The New York Times*, October 11.

BLS (Bureau of Labor Statistics), 1985. *Handbook of Labor Statistics.* Bureau of Labor Statistics, U.S Department of Labor, Bulletin 2217, Washington, D.C., June, pp. 164-171.

BLS, 1987. *Workforce 2000—Work and Workers for the 21st Century.* Johnston, William B., and Packer, Arnold E., Hudson Institute, Indianapolis, Indiana, June.

BLS, 1988a. *Occupational Outlook Handbook, 1988-89.* Bureau of Labor Statistics, U.S. Department of Labor, Washington, D.C., pp. 65-72.

BLS, 1988b. *Occupational Outlook Quarterly.* Volume 32, Number 1, Bureau of Labor Statistics, U.S Department of Labor, Washington, D.C., Spring.

BOC (Bureau of the Census), 1982. *Preliminary Estimates of the Population of the United States, by Age, Sex, and Race: 1970 to 1981.* Current Population Reports, Series P-25, No. 917, Bureau of the Census, Washington, D.C.

BOC, 1984. *Projections of the Population of the United States, by Age, Sex, and Race: 1983 to 2080.* Current Population Reports, Series P-25, No. 952, Bureau of the Census, Washington, D.C.

BOC, 1986. *Projections of the Hispanic Population: 1983 to 2080.* Current Population Reports, Series P-25, No. 995, Bureau of the Census, Washington, D.C.

BOC, 1988a. *Statistical Abstract of the United States 1988*. 108th Edition, Selected Tables, Bureau of the Census, Washington, D.C.

BOC, 1988b. *United States Population Estimates, by Age, Sex, and Race: 1980 to 1987*. Current Population Reports, Series P-25, No. 1022, Bureau of the Census, Washington, D.C.

BRKN (Brookings Institute), 1985. *The State of Graduate Education*. Smith, Bruce L. R. (editor). Brookings Institute, Washington, D.C.

Business Week, 1988. "Human Capital: The Decline of America's Work Force," *Business Week,* September 19, pp. 100-141.

CARN (Carnegie Foundation for the Advancement of Teaching), 1983. *School and College.* Maeroff, Gene I., The Carnegie Foundation for the Advancement of Teaching, Princeton, New Jersey.

CARN, 1984. *High School: A Report on Secondary Education in America.* Boyer, Ernest L., The Carnegie Foundation for the Advancement of Teaching, Harper & Row, New York.

CARN, 1987. *College: The Undergraduate Experience in America.* Boyer, Ernest L., The Carnegie Foundation for the Advancement of Teaching, Harper & Row, New York.

CBMS (Conference Board of the Mathematical Sciences), 1981. *Undergraduate Mathematical Sciences in Universities, Four-Year Colleges and Two-Year Colleges, 1980-81.* Fey, James T., Albers, Donald J., and Fleming, Wendell H., Report of the Survey Committee, Conference Board of the Mathematical Sciences, Washington, D.C.

CBMS, 1982. *Employment of Recent Bachelor's and Master's Graduates in the Mathematical and Computer Sciences.* Botts, Truman, Conference Board of the Mathematical Sciences, Washington, D.C.

CBMS, 1986. *Report of the Committee on American Graduate Mathematics Enrollments.* Conference Board of the Mathematical Sciences, Washington, D.C., unpublished.

CBMS, 1987. *Undergraduate Programs in the Mathematical and Computer Sciences: The 1985-86 Survey.* Albers, Donald J., Anderson, Richard D., and Loftsgaarden, Don O., The Mathematical Association of America, MAA Notes, Number 7, Washington, D.C.

CEEB (College Entrance Examination Board), 1985a. *Academic Preparation in Mathematics: Teaching for Transition from High School to College.* College Entrance Examination Board, New York.

CEEB, 1985b. *Equality and Excellence: The Educational Status of Black Americans.* College Entrance Examination Board, New York.

Bibliography

CEEB, 1987. *National College-Bound Seniors 1987.* College Entrance Examination Board, New York.

CIRP (Cooperative Institutional Research Program), 1987a. *The American Freshman: National Norms for Fall 1987.* Astin, Alexander W., Green, Kenneth C., Korn, William S., and Schalit, Marilynn, Cooperative Institutional Research Program, American Council on Education, University of California, Los Angeles.

CIRP, 1987b. *The American Freshman: Twenty Year Trends.* Astin, Alexander W., Green, Kenneth C., and Korn, William S., Cooperative Institutional Research Program, American Council on Education, University of California, Los Angeles.

Connors, Edward A., 1988. "A Decline in Mathematics Threatens Science—and the U.S.," *The Scientist*, November 28.

CPST (Commission on Professionals in Science and Technology), 1984. *The Science and Engineering Talent Pool: Proceedings.* Scientific Manpower Commission, Washington, D.C.

CPST, 1987. *Scientific Manpower - 1987 and Beyond: Today's Budgets—Tomorrow's Workforce.* Proceedings of a Symposium, October 15, 1986, Commission on Professionals in Science and Technology, Washington, D.C.

CPST, 1988. *Competition for Human Resources in Science and Engineering in the 1990's.* Proceedings of a Symposium, October 11-12, 1987, Commission on Professionals in Science and Technology, Washington, D.C.

DOEd (U.S. Department of Education), 1983. *A Nation At Risk: The Imperative for Educational Reform.* The National Commission on Excellence in Education, U.S. Department of Education, Washington, D.C.

DOEd, 1988. *American Education: Making It Work.* Bennett, William J., U.S. Department of Education, Washington, D.C.

ETS (Educational Testing Service), 1987. *A Summary of Data Collected from Graduate Record Examinations Test Takers During 1985-1986.* Data Summary Report #11, Educational Testing Service, Princeton, New Jersey.

ETS, 1988. *The Mathematics Report Card, Are We Measuring Up?* Dossey, John A., Mullis, Ina V.S., Lindquist, Mary M., and Chambers, Donald L., Educational Testing Service, Princeton, New Jersey.

ETS, 1989. *A World of Differences: An International Assessment of Mathematics and Science.* Lapointe, Archie E., Mead, Nancy A., and Phillips, Gary W., Center for the Assessment of Educational Progress, Educational Testing Service, Princeton, New Jersey.

Fowler, Elizabeth M., 1988. "Careers: A Shortage of Women in Mathematics," *The New York Times*, November 1.

HMSO (Her Majesty's Stationery Office), 1982. *Mathematics Counts.* Cockcroft, W. H., Her Majesty's Stationery Office, London.

IAEEA (International Association for the Evaluation of Education Achievement), 1987. *The Underachieving Curriculum.* McKnight, Curtis C., Crosswhite, F. Joe, Dossey, John A., Kifer, Edward, Swafford, Jane O., Travers, Kenneth J., and Cooney, Thomas J., Stipes Publishing Company, Champaign, Illinois.

IEL (Institute for Educational Leadership), 1985. *All One System: Demographics of Education-Kindergarten through Graduate School.* Hodgkinson, Harold L., Institute for Educational Leadership, Inc., Washington, D.C.

Karmin, Monroe W., 1985. "Jobs of the Future," *U.S. News and World Report*, December 23, p. 40.

MAA (Mathematical Association of America), 1983. "Undergraduate Mathematics in China," Steen, Lynn Arthur, The Newsletter of the Mathematical Association of America, *Focus,* Volume 3, Number 4, September-October 1983, pp. 1-2.

MAA, 1986. *Toward a Lean and Lively Calculus.* Douglas, Ronald G. (editor), The Mathematical Association of America, MAA Notes, Number 6, Washington, D.C.

MAA, 1987a. *Calculus for a New Century: A Pump, Not A Filter.* Steen, Lynn Arthur (editor), The Mathematical Association of America, MAA Notes, Number 8, Washington, D.C.

MAA, 1987b. *Teaching Assistants and Part-Time Instructors: A Challenge.* Case, Bettye Anne (editor), The Mathematical Association of America, Washington, D.C.

Moody, Harry R., 1986. "Education as a Lifelong Process." From *Our Aging Society: Paradox & Promise,* Pifer, Alan, and Bronte, Lydia (editors), Carnegie Corporation of New York, W.W. Norton, New York.

NAS (National Academy of Sciences), 1984. *High Schools and the Changing Workplace: The Employers' View.* Committee on Science, Engineering, and Public Policy, National Academy of Sciences, National Academy Press, Washington, D.C.

NAS, 1986. *Nurturing Science and Engineering Talent: A Symposium, September 29, 1986.* The Franklin Institute, Philadelphia, Pennsylvania, Government-University-Industry Research Roundtable, National Academy of Sciences, unpublished.

NAS,1987a. *Nurturing Science and Engineering Talent—A Discussion Paper.* Government-University-Industry Research Roundtable, National Academy of Sciences, Washington, D.C., unpublished.

Bibliography

NAS, 1987b. *Technology and Employment.* Cyert, Richard M., and Mowery, David C. (editors), Panel on Technology and Employment, Committee on Science, Engineering, and Public Policy, National Academy Press, Washington, D.C.

NAS, 1988. "A Dialogue on Competitiveness," Gomory, Ralph E., and Shapiro, Harold T. *Issues in Science and Technology,* Summer, pp. 36-42, National Academy of Sciences, Washington, D.C.

NASA (National Aeronautics and Space Administration), 1987. *The Future Workforce Conference Proceedings.* Office of Educational Affairs and Office of Productivity Programs, NASA, Government Printing Office, Washington, D.C.

NCES (National Center for Education Statistics), 1984. *Digest of Education Statistics, 1983-84.* Grant, W. Vance, and Snyder, Thomas D., National Center for Education Statistics, U.S. Department of Education, Washington, D.C.

NCES, 1985. *High School and Beyond: a National Longitudinal Study for the 1980's.* West, Jerry, Miller, Wendy, and Diodato, Louis, National Center for Education Statistics, U.S. Department of Education, Washington, D.C.

NCES, 1987a. *Digest of Education Statistics, 1987.* Snyder, Thomas D., National Center for Education Statistics, U.S. Department of Education, Washington, D.C.

NCES, 1987b. *Trends in Bachelors and Higher Degrees, 1975-85.* Carpenter, Judi, National Center for Education Statistics, U.S. Department of Education, Washington, D.C.

NCES, 1988a. *Digest of Education Statistics, 1988.* Snyder, Thomas D., National Center for Education Statistics, U.S. Department of Education, Washington, D.C.

NCES, 1988b. *Trends in Minority Enrollment in Higher Education, Fall 1976-Fall 1986.* Office of Educational Research and Improvement, Survey Report, U.S. Department of Education, Washington, D.C.

NCES, 1989. *College Persistence and Degree Attainment for 1980 High School Graduates: Hazards for Transfers, Stopouts, and Part-Timers.* Office of Educational Research and Improvement, Survey Report, U.S. Department of Education, Washington, D.C.

NRC (National Research Council), 1979. *Climbing the Academic Ladder: Doctoral Women Scientists in Academe.* Committee on the Education and Employment of Women in Science and Engineering, Commission on Human Resources, Office of Science and Technology Policy, National Research Council, National Academy of Sciences, Washington, D.C.

NRC, 1982. *An Assessment of Research-Doctorate Programs in the United States: Mathematical and Physical Sciences.* Jones, Lyle V., Lindzey, Gardner, and Coggeshall, Porter E. (editors), Committee on an Assessment of Quality-Related Characteristics of Research-Doctorate Programs in the United States, Office of Scientific and Engineering Personnel, National Research Council, National Academy Press, Washington, D.C.

NRC, 1983. *Climbing the Ladder: An Update on the Status of Doctoral Women Scientists and Engineers.* Committee on the Education and Employment of Women in Science and Engineering, Office of Scientific and Engineering Personnel, National Research Council, National Academy Press, Washington, D.C.

NRC, 1984. *Renewing U.S. Mathematics: Critical Resource for the Future.* David, Edward E. (chair), Ad Hoc Committee of Resources for the Mathematical Sciences, Commission on Physical Sciences, Mathematics, and Resources, National Research Council, National Academy Press, Washington, D.C.

NRC, 1985. *Indicators of Precollege Education in Science and Mathematics: A Preliminary Review.* Raizen, Senta A., and Jones, Lyle V. (editors), Committee on Indicators of Precollege Science and Mathematics Education, Commission of Behavioral and Social Sciences and Education, National Research Council, National Academy Press, Washington, D.C.

NRC, 1986. *Mathematical Sciences: A Unifying and Dynamic Resource.* Board on Mathematical Sciences, Commission on Physical Sciences, Mathematics, and Resources, National Research Council, National Academy Press, Washington, D.C.

NRC, 1987. *Summary Report 1986: Doctorate Recipients from United States Universities.* Office of Scientific and Engineering Personnel, National Research Council, National Academy Press, Washington, D.C.

NRC, 1989. *Everybody Counts: A Report to the Nation on the Future of Mathematics Education.* Mathematical Sciences Education Board, Board on Mathematical Sciences, and Committee on the Mathematical Sciences in the Year 2000, National Research Council, National Academy Press, Washington, D.C.

NSB (National Science Board), 1985. *Science Indicators: The 1985 Report.* National Science Board, National Science Foundation, Washington, D.C.

NSB, 1986. *Undergraduate Science, Mathematics and Engineering Education.* National Science Board Task Committee on Undergraduate Science and Engineering Education, National Science Foundation, Washington, D.C.

NSB, 1987. *Science & Engineering Indicators—1987.* National Science Board, National Science Foundation, Washington, D.C.

NSF (National Science Foundation), 1982a. *Science and Engineering Degrees: 1950-80. A Source Book.* National Science Foundation, Washington, D.C.

NSF, 1982b. *Science and Engineering Education: Data and Information.* Office of Scientific and Engineering Personnel and Education, National Science Foundation, Washington, D.C.

NSF, 1983. *Science and Engineering Doctorates: 1960-82.* Surveys of Science Resources Series, National Science Foundation, Washington, D.C.

NSF, 1984a. *Academic science/engineering: scientists and engineers, january 1983.* National Science Foundation, 84-309, Washington, D.C.

NSF, 1984b. *Science and Engineering Personnel: A National Overview.* National Science Foundation, Washington, D.C.

NSF, 1985. *Academic science/engineering: graduate enrollment and support, fall 1985.* Surveys of Science Resources Series, National Science Foundation, Washington, D.C.

NSF, 1986a. *Faculty Quick Response Survey, 1986.* Division of Science Resource Studies, Survey Report (unpublished data), National Science Foundation, Washington, D.C.

NSF, 1986b. *Foreign Citizens in U.S. Science and Engineering: History, Status, and Outlook.* Surveys of Science Resources Series, National Science Foundation, Washington, D.C.

NSF, 1986c. *Science and Engineering Research Facilities at Doctorate-Granting Institutions.* National Science Foundation, Washington, D.C.

NSF, 1986d. *Selected Data on Graduate Science/Engineering Students and Postdoctorates: Fall 1985.* Division of Science Resources Studies, National Science Foundation, Washington, D.C.

NSF, 1986e. *Women and Minorities in Science and Engineering.* National Science Foundation, Washington, D.C.

NSF, 1987a. *Characteristics of Recent Science/Engineering Graduates: 1986.* Division of Science Resources Studies, National Science Foundation, Washington, D.C.

NSF, 1987b. *The Science and Engineering Pipeline.* Division of Policy Research and Analysis, National Science Foundation, Washington, D.C.

NSF, 1987c. "Recent-Doctorate Faculty Increase in Engineering and Some Science Fields," *Highlights*, Science Resources Studies, National Science Foundation, Washington, D.C., July.

NSF, 1988a. *Academic science/engineering: graduate enrollment and support, fall 1986.* Surveys of Science Resources Series, 88-307, National Science Foundation, Washington, D.C.

NSF, 1988b. *National Patterns of Science and Technology Resources: 1987.* Surveys of Science Resources Series, 88-305, National Science Foundation, Washington, D.C.

NSF, 1988c. *Science and Engineering Doctorates, 1960-1986.* NSF 88-309, National Science Foundation, Washington, D.C.

NSF, 1988d. *Women and Minorities in Science and Engineering.* National Science Foundation, Washington, D.C.

ORAU (Oak Ridge Associated Universitites), 1985. *Foreign National Scientists and Engineers in the U.S. Labor Force, 1972-1982.* Labor and Policy Studies Program, Manpower Education, Research and Training Division, Oak Ridge Associated Universities, Oak Ridge, Tennessee.

ORAU, 1988. *Estimating Emigration of Foreign-Born Scientists and Engineers in the United States* (Working Paper). Finn, Michael G., and Clark, Sheldon B., Manpower Education, Research and Training Division, Oak Ridge Associated Universities, Oak Ridge, Tennessee.

OTA (Office of Technology Assessment), 1985. *Demographic Trends and the Scientific and Engineering Work Force.* Office of Technology Assessment, U.S. Congress, Washington. D.C.

OTA, 1987. *Preparing for Science and Engineering Careers: Field-Level Profiles* (Staff Paper). Office of Technology Assessment, U.S. Congress, Washington D.C.

OTA, 1988a. *Educating Scientists and Engineers: Grade School to Grad School.* Office of Technology Assessment, U.S. Congress, Washington D.C.

OTA, 1988b. *Elementary and Secondary Education for Science and Engineering.* Office of Technology Assessment, U.S. Congress, Washington D.C.

OTA, 1989. *Higher Education for Science and Engineering: A Background Paper.* Office of Technology Assessment, U.S. Congress, Washington D.C.

Peterson, Ivars, 1988. "Modern Mathematics Is Rich with Provocative Ideas," *Chronicle of Higher Education*, December 7.

RAND (The Rand Corporation), 1984. *Beyond The Commission Reports: The Coming Crisis in Teaching.* Darling-Hammond, Linda, The Rand Corporation, R-3177-RC, Santa Monica, California.

Bibliography

RAND, 1987. *Indicator Systems for Monitoring Mathematics and Science Education.* Shavelson, Richard, McDonnell, Lorraine, Oakes, Jeannie, and Carey, Neil, The Rand Corporation, R-3570-NSF, Santa Monica, California.

RAND, 1988. *The Evolution of Teacher Policy.* Darling-Hammond, Linda, and Berry, Barnett, The Rand Corporation, JRE-01, Santa Monica, California, March.

Richman, Louis S., 1988. "Tomorrow's Jobs: Plentiful, But . . . ," *Fortune* , April 11, pp. 42-56.

ROCK (The Rockefeller Foundation), 1983. *Who Will Do Science?* Berryman, Sue E., The Rockefeller Foundation, A Special Report, November.

RTI (Research Triangle Institute), 1987. *Report of the 1985-86 National Survey of Science and Mathematics Education.* Weiss, Iris R., Research Triangle Institute, RTI/2938/00-FR, Research Triangle Park, North Carolina.

SCC (Science Council of Canada), 1976. *Mathematical Sciences in Canada.* Beltzner, Klaus P., Coleman, A. John, and Edwards, Gordon D., Science Council of Canada, Background Study No. 37, Maracle Press, Oshawa, Canada.

Shogren, Elizabeth, 1988. "Best Job? It's Actuary," Associated Press news article, August.

SREB (Southern Regional Education Board), 1980. *Engineering and High Technology Manpower Shortages: The Connection with Mathematics.* Galambos, Eva C., Southern Regional Education Board, Atlanta, Georgia.

SREB, 1985. *Access to Quality Undergraduate Education.* A Report to the Southern Regional Education Board by Its Commission for Educational Quality, Southern Regional Education Board, Atlanta, Georgia.

TFST (Task Force on Women, Minorities, and the Handicapped in Science and Technology), 1988. *Changing America: The New Face of Science and Engineering* (Interim Report). Oaxaca, Jaime, and Reynolds, Ann W., Task Force on Women, Minorities, and the Handicapped in Science and Technology, Washington, D.C. September.

Vobejda, Barbara, 1989. "Survey of Math Science Skills Puts U.S. at Bottom," *The Washington Post*, February 1.

Appendix Tables

List of Appendix Tables

A2.1	Educational attainment of the civilian labor force (percent distribution)	91
A2.2	Population of 18- to 24-year-olds	91
A2.3	High school dropouts among 18- and 19-year-olds (percent distribution)	92
A2.4	Enrollment rates of 18- to 24-year-olds in institutions of higher education	92
A2.5	Enrollment rates of 25- to 34-year-olds in institutions of higher education	92
A2.6	Bachelor's degrees conferred by institutions of higher education	93
A2.7	Undergraduate enrollment in institutions of higher education	94
A2.8	Persistence rates for 1980 high school graduates	94
A2.9	Anticipated college major (percent distribution)	95
A2.10	Numbers and attainment rates of masters and doctoral degrees for selected fields, 1971 to 1985	96
A3.1	Attitudes of 8th and 12th grade mathematics students toward mathematics, 1981-1982 school year	97
A3.2	Average SAT scores in mathematics, 1970 to 1987	98
A3.3	Average ACT scores in mathematics, 1970 to 1988	98
A3.4	Average NAEP scores in mathematics, 1973 to 1986	99
A3.5	Enrollments in selected mathematics courses in colleges and universities (in thousands)	100
A3.6	Enrollments in undergraduate mathematical sciences departments by type of institution (in thousands)	100
A3.7	Mean number of semester credits completed by bachelor's degree recipients, by major and by course area: 1972 to 1976 and 1980 to 1984	101
A4.1	Number of mathematical sciences degrees awarded, 1950 to 1986	102
A4.2	Comparison of actual versus expected number of mathematical sciences bachelor's degrees	103
A4.3	Anticipated college major and probable career occupation (percent distribution)	103
A4.4	Number of education and mathematics education degrees for selected years	104
A4.5	Full-time graduate students in doctorate-granting institutions for selected fields, 1975 to 1986	105
A4.6	Enrollments in doctorate-granting institutions for mathematical sciences by sex, 1975 to 1986	106
A4.7	Source of major support for full-time mathematical sciences graduate students in doctorate-granting institutions, 1986	107
A4.8	Type of major support for full-time graduate students in doctorate-granting institutions for selected fields, 1986	107
A4.9	1986 enrollments in graduate mathematical sciences programs	108
A4.10	Number of mathematical sciences master's degrees awarded by subfield	109

A4.11	Attainment rates of master's and doctoral mathematics degrees by sex, 1970 to 1986	109
A4.12	Characteristics of new doctorates in mathematical sciences, 1974 to 1986	110
A4.13	Primary sources of support of doctorate recipients in the physical sciences, 1977 and 1986	110
A4.14	Number of doctorate recipients in broadly interpreted mathematical sciences, 1976 to 1986, awarded by U.S. universities	111
A4.15	Total number and distribution of new mathematical sciences doctorates by subfield and sex, 1960 to 1982	112
A4.16	Number of doctorates awarded in selected fields, 1970 to 1985	113
A5.1	Number of employed scientists and engineers	113
A5.2	Selected employment characteristics of scientists and engineers, 1986	113
A5.3	Number of scientists by field and type of employer, 1976 and 1986	114
A5.4	Field of employment for recent (1984/85) mathematics degree recipients, 1986 (percent distribution)	115
A5.5	Primary work activities for recent (1984/85) mathematics degree recipients, 1986 (percent distribution)	115
A5.6	Median annual salaries by field and type of degree of recent (1984/85) graduates, 1986	115
A5.7	Demand and supply of new teachers in elementary and secondary schools, 1970 to 1992 (in thousands)	115
A5.8	Numbers of faculty members by types of institutions for selected years	116
A5.9	Mathematical sciences and computer science enrollments per full-time equivalent (FTE) of faculty	117
A5.10	Age distribution of full-time mathematical sciences faculty in 1985 in four-year colleges and universities	117
A5.11	Age distribution of full-time mathematical sciences faculty in 1985 in two-year colleges	118
A5.12	Employment status of new doctorates awarded by U.S. and Canadian mathematical sciences departments	118
A5.13	Mathematical sciences faculty salaries by type of institution (in 1985 constant dollars)	119

TABLE A2.1 Educational attainment of the civilian labor force (percent distribution)

Years of school	1965	1975	1984	2000 (new jobs)
Less than 4 years of high school	42.5	29.3	19.5	14.0
High school (4 years)	35.5	39.6	40.7	35.0
College (1-3 years)	10.5	15.5	19.0	22.0
College (4+ years)	11.6	15.7	20.9	30.0
Median number of years in school	12.2	12.5	12.8	13.5

SOURCES: Bureau of Labor Statistics (BLS, 1987, p. 98) and Bureau of Labor Statistics (BLS, 1985, p. 164).

TABLE A2.2 Population of 18- to 24-year-olds

Race or ethnic group	1970	1975	1980	1985	1990	1995	2000	2010
Total (in millions)	24.7	28.0	30.3	28.7	25.8	23.7	24.6	27.7
Hispanic origin	a	a	a	2.3	2.4	2.5	2.8	3.6
White	21.5	24.0	25.6	21.6	18.9	16.9	17.2	18.6
Black	2.8	3.5	4.0	4.1	3.8	3.5	3.8	4.6
Other races	0.4	0.5	0.7	0.8	0.8	0.9	1.0	1.1
Total (percent distribution)				100	100	100	100	100
Hispanic origin				8	9	11	11	13
White				75	73	71	70	67
Black				14	15	15	15	17
Other races				3	3	4	4	4

[a] Persons of Hispanic origin were included in white, black, and other races during 1970 to 1980.
SOURCES: Bureau of the Census (BOC, 1982; 1986, p. 14).

TABLE A2.3 High school dropouts among 18- and 19-year-olds (percent distribution)

Race or ethnic group	1970	1975	1980	1985
Total	16.2	16.0	15.7	14.3
Hispanic origin	—	30.1	39.0	30.6
White	14.1	14.7	14.9	13.8
Black	31.2	25.4	21.2	17.3

NOTE: High school dropouts are considered to be those not enrolled in school and not high school graduates.
SOURCE: National Center for Education Statistics (NCES, 1987a, p. 86).

TABLE A2.4 Enrollment rates of 18- to 24-year-olds in institutions of higher education

Race or ethnic group	1970	1975	1980	1985	1986
Enrollment as a percent of 18 to 24 year olds					
Total	25.7	26.3	25.6	27.8	—
Hispanic origin	—	20.4	16.1	16.9	—
White	27.1	26.9	26.2	28.7	—
Black	15.5	20.7	19.2	19.8	—
Enrollment as a percent of high school graduates					
Total	32.7	32.5	31.6	33.7	34.0
Hispanic origin	—	35.5	29.8	26.9	29.4
White	33.2	32.4	31.8	34.4	34.5
Black	26.0	32.0	27.6	26.1	28.6

SOURCES: National Center for Education Statistics (NCES, 1987a, p. 155; 1988b, p. 14).

TABLE A2.5 Enrollment rates of 25- to 34-year-olds in institutions of higher education

Race or ethnic group	1976	1980	1984	1986
Enrollment as a percent of high school graduates				
Total	9.6	8.9	8.6	8.3
Hispanic origin	10.9	9.2	9.9	10.4
White	9.2	8.7	8.4	8.0
Black	11.9	9.6	8.1	7.8

SOURCE: National Center for Education Statistics (NCES, 1988b, p. 14).

TABLE A2.6 Bachelor's degrees conferred by institutions of higher education

Major	1971	1975	1980	1985	1971 to 1985 (% change)
Total	839,730	922,933	929,417	979,477	17
Agriculture and home economics	23,839	34,300	41,213	33,662	41
Business and management	114,865	133,010	185,361	233,351	103
Computer and information sciences	2,388	5,033	11,154	38,878	1,528
Education	176,614	167,015	118,169	88,161	-50
Engineering and engineering technologies	50,046	46,852	68,893	96,105	92
English/letters	64,933	48,534	33,497	34,091	-47
Fine arts	30,394	40,782	40,892	37,936	25
Health sciences	25,190	48,858	63,607	64,513	156
Humanities[a]	44,741	53,112	54,176	65,618	47
Life sciences	35,743	51,741	46,370	38,445	8
Mathematics	24,801	18,181	11,378	15,146	-39
Physical sciences	21,412	20,778	23,410	23,732	11
Social sciences and psychology	193,116	186,153	145,481	131,272	-32
Professional[b]	17,336	39,931	50,660	42,492	145
Other[c]	14,312	28,653	35,156	36,075	152
Total (percent distribution)	100	100	100	100	
Agriculture and home economics	3	4	4	3	
Business and management	14	14	20	24	
Computer and information sciences	0	1	1	4	
Education	21	18	13	9	
Engineering and engineering technologies	6	5	7	10	
English/letters	8	5	4	3	
Fine arts	4	4	4	4	
Health sciences	3	5	7	7	
Humanities[a]	5	6	6	7	
Life sciences	4	6	5	4	
Mathematics	3	2	1	2	
Physical sciences	3	2	3	2	
Social sciences and psychology	23	20	16	13	
Professional[b]	2	4	5	4	
Other[c]	2	3	4	4	

[a] Includes area and ethnic studies, communications, foreign languages, philosophy, religion, and theology.
[b] Includes architecture and environmental design, communication technologies, library and archival sciences, military sciences, parks and recreation, protective services, and public affairs.
[c] Includes law, liberal/general studies, and multi/interdisciplinary studies.
SOURCE: National Center for Education Statistics (NCES, 1987a, p. 190).

TABLE A2.7 Undergraduate enrollment in institutions of higher education

Race or ethnic group	1976	1978	1980	1982	1984	1986
Total (in millions)	9.50	9.80	10.60	10.90	10.60	10.80
Hispanic origin	0.36	0.39	0.44	0.49	0.50	0.57
White	7.80	7.90	8.60	8.70	8.50	8.60
Black	0.95	0.98	1.00	1.00	1.00	1.00
Total (percent distribution)	100	100	100	100	100	100
Hispanic origin	3.7	4.0	4.1	4.5	4.7	5.3
White	82.2	81.4	81.0	80.5	80.0	79.2
Black	10.0	10.0	9.7	9.4	9.4	9.2

NOTE: Figures include nonresident aliens.
SOURCE: National Center for Education Statistics (NCES, 1988b, p. 12).

TABLE A2.8 Persistence rates for 1980 high school graduates

	AY 80-81	Summer 1981	AY 81-82	Summer 1982	AY 82-83	Summer 1983	AY 83-84	From start	From high school
Total	93.3	92.5	87.3	93.3	96.4	95.1	84.2	54.3	15.7
Gender									
Male	94.1	93.1	86.5	93.5	96.4	95.9	85.1	55.7	15.4
Female	92.5	91.9	88.1	93.1	96.4	94.4	83.4	53.0	16.0
Race or ethnicity									
White	93.4	93.1	87.7	93.6	96.7	95.5	84.3	55.6	16.9
Black	91.7	89.9	83.8	90.4	93.0	92.1	81.3	43.5	11.6
Hispanic	90.6	83.0	85.6	92.3	96.6	92.2	80.1	42.3	6.5
Asian	99.4	91.8	90.1	90.8	98.5	93.4	88.6	60.9	27.1
Socioeconomic quartile									
Low quartile	91.7	90.4	79.9	91.3	94.9	86.4	84.2	41.7	6.1
High quartile	95.9	93.5	90.8	93.6	96.9	97.0	84.3	60.4	31.9
Type of college 9/80									
Public 4-year	93.1	92.3	87.6	93.2	96.2	93.8	82.9	52.5	a
Private 4-year	93.6	92.8	86.8	93.3	96.8	97.5	86.6	57.6	a

[a] Not computed.
SOURCE: National Center for Education Statistics (NCES, 1989).

TABLE A2.9 Anticipated college major (percent distribution)

Major	1966	1970	1975	1980	1985
Total	100	100	100	100	100
Agriculture	1.9	2.0	3.9	2.9	2.0
Business	14.3	16.2	18.9	21.3	24.8
Computer science	NA	NA	1.0[a]	2.5	2.3
Education	10.6	11.6	9.9	7.7	7.1
Engineering	9.8	8.6	7.9	11.8	10.7
English	4.4	3.0	1.0	0.9	1.0
Fine arts	8.4	9.2	6.2	5	3.8
Health sciences	5.3	7.4	7.3	9.2	8.9
Humanities	4.7	3.5	2.1	2.1	2.1
Life sciences	3.7	3.5	6.3	3.7	3.4
Mathematical sciences	4.5	3.3	1.1	0.6	0.8
Physical sciences	3.3	2.3	2.7	2.0	1.6
Social sciences	NA	8.9	6.2	4.7	5.2
Other or undecided	29.1	20.5	25.5	25.6	26.3

[a] Data for 1977, which is first year data available.

NOTE: NA means not available.

SOURCE: Cooperative Institutional Research Program (CIRP, 1987b, p. 90).

TABLE A2.10 Numbers and attainment rates of master's and doctoral degrees for selected fields, 1971 to 1985

Field	Degrees awarded for specified period		Attainment rate
	B.S. (1971 to 1983)	M.S. (1973 to 1985)	M.S./B.S. (2-year lag)
Engineering	659,101	219,700	33%
Life sciences	594,785	81,221	14%
Physical sciences	289,783	72,614	25%
Mathematical sciences	212,460	44,764	21%

Field	Degrees awarded for specified period		Attainment rate
	M.S. (1971 to 1980)	Ph.D. (1976 to 1985)	Ph.D./M.S. (5-year lag)
Engineering	158,970	27,035	17%
Life sciences	65,037	34,947	54%
Physical sciences	57,808	32,501	56%
Mathematical sciences	41,399	7,439	18%

Field	Degrees awarded for specified period		Attainment rate
	B.S. (1971 to 1978)	Ph.D. (1978 to 1985)	Ph.D./B.S. (7-year lag)
Engineering	345,122	21,633	6%
Life sciences	374,732	28,158	8%
Physical sciences	171,757	25,729	15%
Mathematical sciences	154,146	5,760	4%

SOURCE: Adapted from National Center for Education Statistics (NCES, 1987a, pp. 190-192).

TABLE A3.1 Attitudes of 8th and 12th grade mathematics students toward mathematics, 1981-1982 school year

Survey statement	Percent giving a high rating	
	8th grade	12th grade
I usually understand what we are talking about in class	71	75
I really want to do well in mathematics	87	91
I feel good when I solve a mathematics problem by myself	78	91
My parents really want me to do well in mathematics	86	89
It does not scare me to have to take mathematics	71	81
Mathematics is easier for me than for most persons	56	76
If I had a choice, I would learn more mathematics	66	76
Mathematics helps me think logically	64	85
There is usually a rule to follow in mathematics	82	67

	Percent of 8th grade students giving a high rating to:				Percent of 12th grade students giving a high rating to:		
	Importance	Ease	Likes		Importance	Ease	Likes
Memorizing	84	36	22	Equations	94	71	71
Measures	83	52	43	Checking	90	78	29
Checking	79	72	25	Memorizing	85	46	18
Equations	78	53	44	Calculators	80	95	85
Decimals	76	56	44	Word problems	78	26	29
Estimating	68	68	49	Function graphs	69	64	37
Charts and graphs	67	74	57	Probability	66	34	37
Ratios and proportions	67	45	35	Charts and graphs	62	80	42
Word problems	66	39	28	Complex numbers	62	52	37
Tables	61	49	37	Derivatives	56	37	34
Geometric figures	60	45	38	Limits	55	42	30
Calculators	54	86	75	Sequences and series	53	43	33
Inequalities	50	35	25	Proofs	52	20	18
Sets	47	57	38	Vectors	48	41	30
Drawing figures	46	55	42	Integrals	46	26	25
All subtopics	66	55	40	All subtopics	66	50	37

SOURCE: As reported in National Science Board (NSB, 1987, pp. 193-194).

TABLE A3.2 Average SAT scores in mathematics, 1970 to 1987

	1970	1975	1980	1985	1987
All students	488	472	466	475	476
Male	509	495	491	499	500
Female	465	449	443	452	453
American Indian	—	420 [a]	426	428	432
Asian-American	—	518 [a]	509	518	521
Black	—	354 [a]	360	376	377
Mexican-American	—	410 [a]	413	426	424
Puerto Rican	—	401 [a]	394	409	400
White	—	493 [a]	482	490	489

[a] Average scores for 1976.

SOURCE: College Entrance Examination Board (CEEB, 1987, pp. iii, v).

TABLE A3.3 Average ACT scores in mathematics, 1970 to 1988

	1970	1975	1980	1985	1986	1987	1988
All students	20.0	17.6	17.4	17.2	17.3	17.2	17.2
Male	21.1	19.3	18.9	18.6	18.8	18.6	18.4
Female	18.8	16.2	16.2	16.0	16.0	16.1	16.1

SOURCES: American College Testing Program (ACT, 1989) and Bureau of the Census (BOC, 1988a, p. 137).

TABLE A3.4 Average NAEP scores in mathematics, 1973 to 1986

	1973	1978	1982	1986
Nine-year-olds				
Total	219.1	218.6	219.0	221.7
Male	217.7	217.4	217.1	221.7
Female	220.4	219.9	220.8	221.7
White	224.9	224.1	224.0	226.9
Black	190.0	192.4	194.9	201.6
Hispanic	202.1	202.9	204.0	205.4
Thirteen-year-olds				
Total	266.0	264.1	268.6	269.0
Male	265.1	263.6	269.2	270.0
Female	266.9	264.7	268.0	268.0
White	273.7	271.6	274.4	273.6
Black	227.7	229.6	240.4	249.2
Hispanic	238.8	238.0	252.4	254.3
Seventeen-year-olds				
Total	304.4	300.4	298.5	302.0
Male	308.5	303.8	301.5	304.7
Female	300.6	297.1	295.6	299.4
White	310.1	305.9	303.7	307.5
Black	269.8	268.4	271.8	278.6
Hispanic	277.2	276.3	276.7	283.1

NOTE: The NAEP has summarized trends in average mathematics proficiency on a common scale ranging from 0 to 500. This is a composite of the student performance on the five content areas. It takes the form of the score on a hypothetical 500-item test composed of questions reflecting the proportional weighting of the subareas. Those average scores are given above, including the ones extrapolated for 1973.

SOURCE: Educational Testing Service (ETS, 1988).

TABLE A3.5 Enrollments in selected mathematics courses in colleges and universities (in thousands)

Subject	1965	1970	1975	1980	1985	1965-1985 (% increase)
Arithmetic and general math	29	23	32	63	45	55
High school algebra and geometry	60	78	109	179	202	237
College algebra, trigonometry	262	301	259	345	352	34
Total math enrollment	1,005	1,215	1,252	1,525	1,619	61

SOURCE: Conference Board of the Mathematical Sciences (CBMS, 1987, p. 21).

TABLE A3.6 Enrollments in undergraduate mathematical sciences departments by type of institution (in thousands)

	1960	1965 [a]	1970	1975	1980	1985
Two-year colleges						
Remedial	NA	109	191	346	441	482
Pre-calculus	NA	92	124	149	175	182
Calculus	NA	46	69	76	91	103
Other	NA	91	171	266	218	133
Computer/data processing	NA	5	13	10	95	98
Statistics	NA	5	16	27	28	36
Total	NA	348	584	874	1,048	1,034
Four-year colleges and universities						
Remedial	96	89	101	141	242	251
Pre-calculus	349	468	538	555	602	593
Calculus	180	315	414	450	590	637
Advanced	92	133	162	106	91	138
Total	717	1,005	1,215	1,252	1,525	1,619
Total undergraduate enrollments in mathematical sciences departments						
Remedial	NA	198	292	487	683	733
Pre-calculus[b]	NA	661	862	1,007	1,118	1,042
Calculus	NA	361	483	526	681	740
Advanced	NA	133	162	106	91	138
Total	NA	1,353	1,799	2,126	2,573	2,653
Undergraduate enrollments in statistics						
Elementary	4	11	57	99	104	144
Advanced	16	32	35	42	43	64
Total	20	43	92	141	147	208

[a] 1966 data used for two-year colleges, since 1965 data are not available.
[b] Includes other, computer/data processing, and statistics at two-year colleges.
NA Data not reported or not applicable.
SOURCE: Conference Board of the Mathematical Sciences (CBMS, 1987, pp. 18, 31, 120).

TABLE A3.7 Mean number of semester credits completed by bachelor's degree recipients, by major and by course area: 1972 to 1976 and 1980 to 1984

Selected college majors	Course areas									
	Total	Bus.	Comp. sci.	Educ.	Eng.	Math.	Bio. sci.	Phys. sci.	Social sci.	Other
1972 to 1976[a]										
Mean, all majors	124.0	7.8	1.0	9.7	2.3	7.4	7.6	9.0	30.3	48.8
Business and management	124.4	41.2	2.3	0.5	0.4	10.2	2.5	4.8	30.4	32.0
Computer science	133.3	6.6	33.5	0.4	5.3	22.4	1.9	7.8	20.6	34.8
Education	126.4	0.9	0.3	40.2	NA	5.0	5.5	4.3	23.9	46.4
Engineering	134.8	1.6	2.0	0.1	50.0	18.2	1.3	20.5	14.0	27.1
English	117.8	0.5	0.1	7.8	0.1	3.2	3.4	3.4	24.2	75.2
Fine arts	124.9	0.3	0.1	6.6	NA	1.3	2.5	2.1	13.6	98.4
Life sciences	122.2	0.4	0.8	1.7	NA	8.4	35.6	26.2	17.8	31.3
Physical sciences	122.7	0.8	1.4	0.9	1.9	16.2	9.6	49.5	13.1	29.2
Psychology	119.1	2.0	0.5	5.9	0.3	5.5	6.2	5.9	56.0	36.9
Social sciences	120.6	3.4	0.4	3.3	0.4	5.3	3.2	4.3	60.3	40.1
1980 to 1984[b]										
Mean, all majors	123.5	12.8	3.3	6.2	4.6	8.4	5.3	8.1	27.5	47.2
Business and management	122.8	41.2	4.5	0.6	1.1	8.9	2.2	3.9	27.5	32.7
Computer science	129.3	11.8	27.9	0.3	4.7	21.3	1.8	8.5	19.0	33.9
Education	127.4	0.7	0.3	45.5	0.1	4.4	4.4	3.8	20.8	47.3
Engineering	132.3	1.0	2.3	0.8	52.5	16.2	1.1	20.2	12.3	25.9
English	114.8	1.7	1.5	6.9	NA	2.2	2.1	4.7	21.4	74.4
Fine arts	120.5	1.7	0.6	5.1	NA	1.7	2.7	1.5	14.1	93.1
Life sciences	121.9	0.7	1.5	1.9	0.2	10.1	33.5	22.6	18.1	33.3
Physical sciences	124.3	0.2	4.9	0.1	2.0	14.1	12.9	48.7	11.6	30.0
Psychology	120.7	3.0	2.7	2.1	NA	6.5	5.8	4.2	55.2	41.2
Social sciences	119.2	6.0	1.4	1.0	0.5	5.4	4.4	5.1	52.0	43.3

[a] Sample survey based on 1972 high school seniors who completed bachelor's degrees by 1976.
[b] Sample survey based on 1980 high school seniors who completed bachelor's degrees by 1984.
NA Data not reported or not applicable.
NOTE: Because of rounding, details may not add to totals.
SOURCES: National Center for Education Statistics (NCES, 1985; 1987a, p. 220).

TABLE A4.1 Number of mathematical sciences degrees awarded, 1950 to 1986

	Bachelor's degrees			Master's degrees			Doctoral degrees		
	Total	Men	Women	Total	Men	Women	Total	Men	Women
1950	6,382	4,942	1,440	974	784	190	160	151	9
1952	4,696	3,374	1,322	802	663	139	206	195	11
1954	4,078	2,717	1,361	706	579	127	227	213	14
1956	4,646	3,128	1,518	898	719	179	235	225	10
1958	6,905	4,943	1,962	1,234	994	240	247	232	15
1960	11,399	8,293	3,106	1,757	1,422	335	303	285	18
1962	14,570	10,331	4,239	2,680	2,179	501	396	372	24
1964	18,624	12,656	5,968	3,597	2,911	686	596	567	29
1966	19,977	13,326	6,651	4,769	3,769	1,000	782	725	57
1968	23,513	14,782	8,731	5,527	4,199	1,328	947	895	52
1970	27,442	17,177	10,265	5,636	3,966	1,670	1,236	1,140	96
1971	24,801	15,369	9,432	5,191	3,673	1,518	1,199	1,106	93
1972	23,713	14,454	9,259	5,198	3,655	1,543	1,128	1,039	89
1973	23,067	13,796	9,271	5,028	3,525	1,503	1,068	966	102
1974	21,635	12,791	8,844	4,834	3,337	1,497	1,031	931	100
1975	18,181	10,586	7,595	4,327	2,905	1,422	975	865	110
1976	15,984	9,475	6,509	3,857	2,547	1,310	856	762	94
1977	14,196	8,303	5,893	3,695	2,396	1,299	823	714	109
1978	12,569	7,398	5,171	3,373	2,228	1,145	805	681	124
1979	11,806	6,899	4,907	3,036	1,985	1,051	730	608	122
1980	11,378	6,562	4,816	2,860	1,828	1,032	724	624	100
1981	11,078	6,342	4,736	2,567	1,692	875	728	614	114
1982	11,599	6,593	5,006	2,727	1,821	906	681	587	94
1983	12,453	6,995	5,458	2,837	1,858	979	698	582	116
1984	13,211	7,366	5,845	2,741	1,791	950	695	569	126
1985	15,146	8,164	6,982	2,882	1,874	1,008	699	590	109
1986	16,306	8,725	7,581	3,159	2,047	1,112	742	618	124

SOURCE: National Center for Education Statistics (NCES, 1988a, p. 237).

Appendix Tables

TABLE A4.2 Comparison of actual versus expected number of mathematical sciences bachelor's degrees

	First year full-time enrollments (in thousands)	Anticipated major mathematics/statsistics (percent)	Expected number of degrees (4 years later)	Actual number of degrees (4 years later)	Ratio, actual/ expected
1967	1,335	4.2	56,070	24,801	0.44
1968	1,471	4.0	58,840	23,713	0.40
1969	1,525	3.5	53,375	23,067	0.43
1970	1,587	3.3	52,371	21,635	0.41
1971	1,606	2.7	43,362	18,181	0.42
1972	1,574	2.2	34,628	15,984	0.46
1973	1,607	1.7	27,319	14,196	0.52
1974	1,673	1.4	23,422	12,569	0.54
1975	1,763	1.1	19,393	11,806	0.61
1976	1,663	1.0	16,630	11,378	0.68
1977	1,681	0.8	13,448	11,078	0.82
1978	1,651	0.9	14,859	11,599	0.78
1979	1,706	0.6	10,236	12,453	1.22
1980	1,749	0.6	10,494	13,211	1.26
1981	1,738	0.6	10,428	15,146	1.45
1982	1,688	0.6	10,128	16,306	1.61
1983	1,678	0.8	13,424	—	—
1984	1,613	0.8	12,904	—	—
1985	1,602	0.8	12,816	—	—

SOURCES: Adapted from Cooperative Institutional Research Program (CIRP, 1987b, p. 90) and National Center for Education Statistics (NCES, 1987a, pp. 130, 207).

TABLE A4.3 Anticipated college major and probable career occupation (percent distribution)

	1966	1970	1975	1980	1985	1987
Anticipated college major						
Education	10.6	11.6	9.9	7.7	7.1	8.9
Mathematics	4.5	3.3	1.1	0.6	0.8	0.6
Probable career occupation						
Elementary education	7.6	8.0	3.0	3.8	3.8	5.0
Secondary education	14.1	11.3	3.5	2.2	2.4	3.1

SOURCES: Cooperative Institutional Research Program (CIRP, 1987a; 1987b, pp. 90, 94).

TABLE A4.4 Number of education and mathematics education degrees for selected years

	1971	1981	1983	1984	1985	1986
Bachelor's degrees						
Education, total	176,614	108,309	97,991	92,382	88,161	87,221
Mathematics education	2,217	798	672	775	1,027	1,259
Percentage of total education degrees	1.3	0.7	0.7	0.8	1.2	1.4
Master's degrees						
Education, total	88,952	98,938	84,853	77,187	76,137	76,353
Mathematics education		372	439	416	426	444
Percentage of total education degrees		0.4	0.5	0.5	0.6	0.6
Doctoral degrees						
Education, total	6,403	7,900	7,551	7,473	7,151	7,110
Mathematics education		30	24	32	29	32
Percentage of total education degrees		0.4	0.3	0.4	0.4	0.5

SOURCES: National Center for Education Statistics (NCES, 1984; 1987a) and unpublished data.

TABLE A4.5 Full-time graduate students in doctorate-granting institutions for selected fields, 1975 to 1986

	1975	1976	1977	1978	1979	1980	1981	1982	1983	1984	1985	1986
Mathematical sciences	10,039	10,157	9,814	9,296	9,136	9,368	9,680	10,174	10,312	10,613	11,168	11,767
U.S. citizens	8,176	7,996	7,433	6,755	6,337	6,213	6,159	6,453	6,232	6,266	6,516	6,805
Foreign students	1,863	2,161	2,381	2,541	2,799	3,155	3,521	3,721	4,080	4,347	4,652	4,962
Physical sciences	21,274	21,582	21,747	21,497	21,797	22,254	22,600	23,330	24,492	25,149	25,967	27,074
U.S. citizens	16,883	17,155	17,174	16,684	16,621	16,668	16,523	17,038	17,451	17,720	17,700	17,922
Foreign students	4,391	4,427	4,573	4,813	5,176	5,586	6,077	6,292	7,041	7,429	8,267	9,152
Engineering	37,083	36,434	36,675	37,057	39,344	41,939	44,817	48,687	53,516	54,757	55,940	59,925
U.S. citizens	25,196	24,161	23,346	22,657	23,133	24,436	25,616	27,875	31,110	31,962	32,203	33,298
Foreign students	11,887	12,273	13,329	14,400	16,211	17,503	19,201	20,812	22,406	22,795	23,737	26,627
Total, science & Engineering	210,321	214,089	217,453	216,608	223,409	230,535	234,194	236,939	243,540	246,718	251,147	259,980
U.S. citizens	177,193	179,689	180,597	176,647	178,659	181,864	181,596	181,637	183,642	185,283	184,753	187,171
Foreign students	33,128	34,400	36,856	39,961	44,750	48,671	52,598	55,302	598,98	61,435	66,394	72,809
Percent distribution												
Mathematical Sciences	100	100	100	100	100	100	100	100	100	100	100	100
U.S. citizens	81	79	76	73	69	66	64	63	60	59	58	58
Foreign students	19	21	24	27	31	34	36	37	40	41	42	42
Physical sciences	100	100	100	100	100	100	100	100	100	100	100	100
U.S. citizens	79	79	79	78	76	75	73	73	71	70	68	66
Foreign students	21	21	21	22	24	25	27	27	29	30	32	34
Engineering	100	100	100	100	100	100	100	100	100	100	100	100
U.S. citizens	68	66	64	61	59	58	57	57	58	58	58	56
Foreign students	32	34	36	39	41	42	43	43	42	42	42	44
Total, science & engineering	100	100	100	100	100	100	100	100	100	100	100	100
U.S. citizens	84	84	83	82	80	79	78	77	75	75	74	72
Foreign students	16	16	17	18	20	21	22	23	25	25	26	28

SOURCE: National Science Foundation (NSF, 1988a, p. 81).

TABLE A4.6 Enrollments in doctorate-granting institutions for mathematical sciences by sex, 1975 to 1986

	1975	1976	1977	1978	1979	1980	1981	1982	1983	1984	1985	1986
Full-time enrollments												
Mathematical sciences	10,039	10,281	9,814	9,296	9,136	9,368	9,680	10,174	10,312	10,613	11,168	11,767
Men	7,918	8,034	7,681	7,271	7,140	7,228	7,322	7,622	7,646	7,876	8,198	8,599
Women	2,121	2,247	2,133	2,025	1,996	2,140	2,358	2,552	2,666	2,737	2,970	3,168
Percent distribution	100	100	100	100	100	100	100	100	100	100	100	100
Men	79	78	78	78	78	77	76	75	74	74	73	73
Women	21	22	22	22	22	23	24	25	26	26	27	27
Part-time enrollments												
Mathematical Sciences	4,352	4,470	4,080	3,898	4,044	4,257	4,324	4,584	4,560	4,454	4,100	3,911
Men	NA	NA	2,836	2,681	2,751	2,940	2,952	3,028	3,001	2,957	2,625	2,477
Women	NA	NA	1,244	1,217	1,293	1,317	1,372	1,556	1,559	1,497	1,475	1,434
Percent distribution	100	100	100	100	100	100	100	100	100	100	100	100
Men	NA	NA	70	69	68	69	68	66	66	66	64	63
Women	NA	NA	30	31	32	31	32	34	34	34	36	37

NA Not available.
SOURCE: National Science Foundation (NSF, 1988a, pp. 80-82).

TABLE A4.7 Source of major support for full-time mathematical sciences graduate students in doctorate-granting institutions, 1986

Source of major support	Men	Women	Total
Total, all sources	8,599	3,168	11,767
Federal	808	184	992
Institutional	5,911	2,271	8,182
Other outside support	484	103	587
Self support	1,396	610	2,006
Percent distribution			
Total, all sources	100	100	100
Federal	9	6	8
Institutional	69	72	70
Other outside support	6	3	5
Self support	16	19	17

SOURCE: National Science Foundation (NSF, 1988a, p. 49).

TABLE A4.8 Type of major support for full-time graduate students in doctorate-granting institutions for selected fields, 1986

Type of major support	Mathematical sciences	Physical sciences	Engineering	Total, sciences	Total, science and engineering
Total	11,767	27,074	59,925	200,055	259,980
Fellowships	871	1,899	4,929	18,516	23,445
Traineeships	136	477	852	13,048	13,900
Research assistantships	1,012	10,847	20,407	45,153	65,560
Teaching assistantships	6,897	11,329	10,973	49,954	60,927
Other types of support	2,851	2,522	22,764	73,384	96,148
Total (percent distribution)	100	100	100	100	100
Fellowships	7	7	8	9	9
Traineeships	1	2	1	7	5
Research assistantships	9	40	34	23	25
Teaching assistantships	59	42	18	25	23
Other types of support	24	9	38	37	37

SOURCE: National Science Foundation (NSF, 1988a, p. 50).

TABLE A4.9 1986 enrollments in graduate mathematical sciences programs

	U.S. Citizens						Other/	
	White	Black	Hispanic	Asian	Unknown	Total U.S.	Foreign	Total
Doctorate granting	8,198	301	248	571	857	10,175	5,503	15,678
Full-time	5,587	191	168	416	443	6,805	4,962	11,767
Part-time	2,611	110	80	155	414	3,370	541	3,911
Master's granting	1,277	147	57	156	659	2,296	405	2,701
Full-time	259	51	14	53	81	458	231	689
Part-time	1,018	96	43	103	578	1,838	174	2,012
Total	9,475	448	305	727	1,516	12,471	5,908	18,379
Full-time	5,846	242	182	469	524	7,263	5,193	12,456
Part-time	3,629	206	123	258	992	5,208	715	5,923
Percent distribution, U.S. citizens								
Doctorate granting	88.0	3.2	2.7	6.1	—	100		
Full-time	87.8	3.0	2.6	6.5	—	100		
Part-time	88.3	3.7	2.7	5.2	—	100		
Master's granting	78.0	9.0	3.5	9.5	—	100		
Full-time	68.7	13.5	3.7	14.1	—	100		
Part-time	80.8	7.6	3.4	8.2	—	100		
Total	86.5	4.1	2.8	6.6	—	100		
Full-time	86.7	3.6	2.7	7.0	—	100		
Part-time	86.1	4.9	2.9	6.1	—	100		

SOURCE: National Science Foundation (NSF, 1988a, pp. 51, 53, 57, 59, 135, 175).

TABLE A4.10 Number of mathematical sciences master's degrees awarded by subfield

	1981	1983	1984	1985	1986	Percent dist., 1986
Total	2,567	2,837	2,741	2,882	3,159	100
Mathematics, general	1,890	1,924	1,846	1,892	2,057	65
Actuarial sciences	NA	27	18	23	28	1
Applied mathematics	179	259	253	284	367	12
Pure mathematics	NA	21	25	27	23	1
Statistics	467	459	471	490	478	15
Mathematics, other	31	147	128	166	206	6

SOURCE: National Center for Education Statistics (NCES, 1984; 1987a, p. 187) and unpublished data.

TABLE A4.11 Attainment rates of master's and doctoral mathematics degrees by sex, 1970 to 1986

Year[a]	Percentage of master's/bachelor's degrees (2-year lag)			Percentage of doctoral/master's degrees (5-year lag)			Percentage of doctoral/bachelor's degrees (7-year lag)		
	Total	Men	Women	Total	Men	Women	Total	Men	Women
1972	19	21	15	—	—	—	—	—	—
1973	20	23	16	—	—	—	—	—	—
1974	20	23	16	—	—	—	—	—	—
1975	19	21	15	—	—	—	—	—	—
1976	18	20	15	—	—	—	—	—	—
1977	20	23	17	16	20	7	3	4	1
1978	21	24	18	16	19	8	3	4	1
1979	21	24	18	15	18	8	3	4	1
1980	23	25	20	17	21	7	3	5	1
1981	22	25	18	19	24	9	3	5	1
1982	24	28	19	18	24	7	4	6	1
1983	26	29	21	21	26	10	4	6	2
1984	24	27	19	23	29	12	5	7	2
1985	23	27	18	24	32	11	6	8	2
1986	24	28	19	29	37	14	6	9	3
15-year average	21	24	17	—	—	—	—	—	—
10-year average	—	—	—	19	24	9	4	5	1.5

[a] Refers to the year in which higher-level degrees were awarded; the lower-level degrees were awarded 2, 5, or 7 years earlier.

SOURCE: Adapted from Table A4.1.

TABLE A4.12 Characteristics of new doctorates in mathematical sciences, 1974 to 1986

	1974	1976	1978	1980	1982	1984	1986
Total number	1,211	1,003	838	744	720	698	730
Percent men	90.5	88.7	85.7	87.2	86.7	83.5	83.4
Percent women	9.5	11.3	14.3	12.8	13.3	16.5	16.6
Percent							
American Indian	0.2	0.0	0.1	0.0	0.1	0.5	0.2
Asian	13.3	12.0	13.4	15.6	16.4	21.5	24.0
Black	1.9	0.8	1.9	2.0	1.8	1.2	1.8
Hispanic	0.7	1.5	3.2	2.3	3.6	5.8	6.0
White	84.0	85.7	81.3	80.1	78.1	71.1	68.0
U.S. citizens	72.3	74.6	73.9	69.9	63.6	58.3	50.3
Permanent visas	5.9	5.5	5.6	8.3	5.7	5.2	4.9
Temporary visas	18.5	18.2	18.5	18.7	26.7	33.2	37.3

SOURCE: National Research Council (NRC, 1987, p. 66).

TABLE A4.13 Primary sources of support of doctorate recipients in the physical sciences, 1977 and 1986

	Personal		University		Federal		Other	
	1977	1986	1977	1986	1977	1986	1977	1986
Total, all fields	36.1	42.1	41.9	44.8	16.1	7.2	5.8	6.0
Physics and astronomy	9.8	7.5	75.8	84.0	10.2	4.4	4.2	4.1
Chemistry	9.8	10.3	74.6	81.4	11.7	5.2	3.9	3.1
Earth/atmospheric/marine sciences	17.3	18.9	56.4	70.9	19.4	6.1	7.0	4.1
Mathematics	17.7	14.0	64.1	74.2	11.8	5.3	6.4	6.5
Computer science	45.0	25.6	55.0	58.2	0.0	4.1	0.0	12.1

NOTE: The primary support of mathematics doctoral students is by their universities, and principally as graduate teaching assistants. The above chart from the 1986 NRC Survey Report shows the changes over 10 years and comparisons with the other physical sciences. (Numbers are percent of total with given source of support.)
SOURCE: National Research Council (NRC, 1987, p. 28).

TABLE A4.14 Number of doctorate recipients in broadly interpreted mathematical sciences, 1976 to 1986, awarded by U.S. universities

	1976	1977	1978	1979	1980	1981	1982	1983	1984	1985	1986
Total, mathematics	696	664	605	563	569	535	533	540	496	540	576
Applied mathematics	105	113	108	111	102	118	108	125	108	116	136
Algebra	116	88	87	88	78	56	60	55	65	55	46
Analysis and functional analysis	141	153	118	111	91	105	98	76	71	83	81
Geometry	23	26	22	25	35	29	32	44	27	35	38
Logic	34	17	24	21	24	18	17	21	25	30	23
Number theory	26	32	18	17	28	24	28	19	27	18	20
Probability (AMS Survey)	42	33	33	27	30	22	25	22	19	39	26
Topology	72	70	56	61	57	55	45	44	42	35	34
Mathematics, general	94	88	92	80	83	77	84	86	78	85	125
Mathematics, other	43	44	47	22	41	31	36	48	34	44	47
Total, computing	267	255	252	313	293	319	303	381	364	381	486
Computing theory and practice[a]	148	101	55	25	13	16	11	12	13	15	10
Computer and information science	—	31	121	210	218	232	220	286	295	310	399
Computer engineering	119	123	76	78	62	71	72	83	56	56	77
Mathematical statistics[a,b]	123	126	135	138	121	141	140	129	162	111	115
Total, statistics	234	242	249	227	218	246	266	250	284	247	238
Biometrics and biostatistics	46	52	45	44	42	48	59	45	49	40	30
Social sciences statistics	35	35	46	23	33	40	43	47	39	60	65
Econometrics	30	29	23	22	22	17	24	21	27	27	25
Business statistics	—	—	—	—	—	—	—	8	7	9	3
Total, operations research	118	118	127	110	104	116	94	102	123	121	129
Mathematics operations research[a]	36	42	43	43	41	36	36	20	27	22	29
Engineering operations research	82	76	84	67	63	80	58	44	50	54	54
Business operations research	—	—	—	—	—	—	—	38	46	45	46
Total, mathematics education	96	98	57	85	74	62	50	62	64	65	72

[a] Reported under "Mathematics" in National Research Council (1987).
[b] These numbers are the NRC counts minus the American Mathematical Society counts for probability given above. It is noted that the AMS counts of doctorates in statistics are considerably higher than these NRC counts. The NRC counts in probability and statistics are approximately the AMS counts for statistics alone.
SOURCES: National Research Council (NRC, 1987, pp. 60-63) and American Mathematical Society (AMS, 1976 to 1988).

TABLE A4.15 Total number and distribution of new mathematical sciences doctorates by subfield and sex, 1960 to 1982

	Number of new doctorates			Distribution by sex (percent)		
	Male (%)	Female (%)	Total	Male	Female	Total
Mathematical sciences, total	18,646 (91)	1,863 (9)	20,509	100	100	100
Algebra	2,135 (86)	338 (14)	2,473	11	18	12
Analysis/functional analysis[a]	3,538 (93)	273 (7)	3,811	19	15	19
Geometry	559 (91)	54 (9)	613	3	3	3
Logic	489 (91)	46 (9)	535	3	2	3
Number theory	468 (88)	63 (12)	531	3	3	3
Probability/math statistics[b]	2,532 (89)	310 (11)	2,842	14	17	14
Topology	1,668 (92)	136 (8)	1,804	9	7	9
Topological algebra[c]	116 (98)	2 (2)	118	1	—	1
Computing theory	1,534 (94)	106 (6)	1,640	8	6	8
Computer science[d]	922 (89)	111 (11)	1,033	5	6	5
Operations research[e]	326 (91)	33 (9)	359	2	2	2
Applied mathematics	2,203 (94)	136 (6)	2,339	12	7	11
Mathematics, general	1,358 (89)	162 (11)	1,520	7	9	7
Mathematics, other	798 (90)	93 (10)	891	4	5	4

[a] This subfield was specified as "Analysis" through 1967; "Functional Analysis" was added in 1968.
[b] "Math Statistics" was deleted from the taxonomy in 1969 but was restored in 1972.
[c] This subfield was deleted in 1967.
[d] This subfield was introduced in 1977.
[e] This subfield was introduced in 1973.
SOURCE: National Science Foundation (NSF, 1983, pp. 18-19, 22-23, 26-27).

TABLE A4.16 Number of doctorates awarded in selected fields, 1970 to 1985

Field	1970	1975	1980	1985	Percent change, 1970 to 1985
Biological sciences	3,361	3,497	3,803	3,771	12
Chemistry	2,238	1,776	1,538	1,837	-18
Mathematical sciences	1,225	1,147	744	688	-44
Physics and astronomy	1,655	1,300	983	1,080	-35
Engineering	4,434	3,002	2,479	3,167	-8

SOURCE: National Science Foundation (NSF, 1988c, pp. 1-5).

TABLE A5.1 Number of employed scientists and engineers

	1976	1980	1986	Percent increase, 1976 to 1986
All science/engineering fields	2,331,200	2,860,400	4,626,500	98
Physical scientists	188,900	215,200	288,400	53
Mathematical scientists	48,600	64,300	131,000	170
Computer specialists	119,000	207,800	562,600	373
Engineers	1,371,700	1,675,900	2,440,100	78
Other[a]	603,100	697,200	1,204,400	100

[a] Includes environmental, life, and social scientists and psychologists.
SOURCE: National Science Board (NSB, 1987, pp. 230-231).

TABLE A5.2 Selected employment characteristics of scientists and engineers, 1986

	Mathematical Scientists			Scientists	Engineers	Total S/E
	Total	Mathematics	Statistics			
Labor force participation rate	94.6	94.5	95.2	95.3	93.8	94.5
Unemployment rate	1.3	1.1	2.4	1.9	1.2	1.5
S/E employment rate	79.3	77.7	87.8	76.7	91.9	84.7
Underemployment rate[a]	3.3	3.4	3.0	4.3	1.0	2.6
Underutilization rate[b]	4.6	4.5	5.3	6.1	2.2	4.1

[a] Those who are involuntarily working either in non-S/E jobs or part-time as a percent of total employment.
[b] Percent of total who are either unemployed or underemployed.
SOURCE: National Science Board (NSB, 1987, p. 225).

TABLE A5.3 Number of scientists by field and type of employer, 1976 and 1986

	Total		Business and industry		Educational institutions		Federal government		Other	
	1976	1986	1976	1986	1976	1986	1976	1986	1976	1986
Total, all S/E fields	2,122,100	3,919,900	1,312,500	2,589,300	267,800	572,700	211,100	334,200	330,700	423,700
Total, scientists	843,700	1,676,400	357,900	797,900	230,200	481,800	105,200	153,500	150,400	243,200
Physical scientists	154,900	264,900	86,800	146,700	30,600	68700	20,500	28,600	17,000	20,900
Mathematical scientists	43,800	103,900	12,100	35,600	19,500	52,800	9,300	10700	2,900	4,800
Computer specialists	116,000	437,200	85,800	341,300	6,000	32,500	8,700	32,100	15,500	31,300
Environmental scientists	46,600	97,300	25,800	55,500	5,000	16,500	9,300	16,800	6,500	8,500
Life scientists	198,200	340,500	64,100	102,800	59,600	136,600	37,300	40,200	37,200	60,900
Psychologists	103,700	172,800	21,000	39,700	41,700	72,800	5,000	5,400	36,000	54,900
Social scientists	180,500	269,800	62,300	76,200	67,800	102,000	15,100	19,700	35,300	71,900
Total, engineering	1,278,300	2,243,500	954,600	1,791,400	37,700	90,900	105,900	180,700	180,100	180,500
Percent distribution										
Total, all S/E fields	100	100	62	66	13	15	10	9	16	11
Total, scientists	100	100	42	48	27	29	12	9	18	15
Physical scientists	100	100	56	55	20	26	13	11	11	8
Mathematical scientists	100	100	28	34	45	51	21	10	7	5
Computer specialists	100	100	74	78	5	7	8	7	13	7
Environmental scientists	100	100	55	57	11	17	20	17	14	9
Life scientists	100	100	32	30	30	40	19	12	19	18
Psychologists	100	100	20	23	40	42	5	3	35	32
Social scientists	100	100	35	28	38	38	8	7	20	27
Total, engineering	100	100	75	80	3	4	8	8	14	8

SOURCE: National Science Board (NSB, 1987, p. 219).

TABLE A5.4 Field of employment for recent (1984/85) mathematics degree recipients, 1986 (percent distribution)

Field	Bachelor's	Master's
Mathematics and statistics	42	61
Computer science	40	15
Engineering	14	17
Economics	2	2
Psychology	—	4

SOURCE: Consolidated from National Science Foundation (NSF, 1987a).

TABLE A5.5 Primary work activities for recent (1984/85) mathematics degree recipients, 1986 (percent distribution)

Primary work activities	Bachelor's	Master's
Research and development	16	17
Management and administration	3	15
Teaching	24	34
Production and inspection	6	6
Reporting/statistical/computing activities	34	15
Other	15	11

SOURCE: Consolidated from National Science Foundation (NSF, 1987a).

TABLE A5.6 Median annual salaries by field and type of degree of recent (1984/85) graduates, 1986

S/E field of degree	Bachelor's	Master's
Engineering	$30,000	$36,000
Computer science	$28,000	$36,600
Total, science and engineering	$25,000	$32,500
Mathematics and statistics	$24,100	$31,500
Physical science	$21,400	$30,000
Total, science	$21,000	$29,000
Environmental science	$20,000	$27,000
Social science	$20,000	$24,600
Life science	$17,000	$22,000
Psychology	$17,000	$23,100

SOURCE: National Science Foundation (NSF, 1987a).

TABLE A5.7 Demand and supply of new teachers in elementary and secondary schools, 1970 to 1992 (in thousands)

	Demand[a]			Supply,[b]	
	Elementary Schools	Secondary Schools	Total	Total	Balance
1970	115	93	208	284	76
1975	101	85	186	238	52
1980	69	58	127	144	17
1985	95	62	157	146	-11
1986	114	56	170	144	-26
1987	114	46	160	142	-18
1988	126	38	164	139	-25
1989	126	47	173	139	-34
1990	131	52	183	139	-44
1991	129	66	195	138	-57
1992	129	80	209	137	-72

[a]Demand: Data for 1970 to 1982 are actual new hires. Data for 1983 to 1992 are the intermediate value of three alternate projections.
[b]Supply: Data for 1984 to 1992 are the intermediate values of three alternate projections.
SOURCE: U.S. Department of Education as reported in National Science Board (NSB, 1987, p. 186).

TABLE A5.8 Numbers of faculty members by types of institutions for selected years

	1970	1975	1980	1985
University				
Mathematics				
Full-time faculty	6,235	5,405	5,605	5,333
Part-time faculty	615	699	1,038	1,044
Statistics				
Full-time faculty	700	732	610	662
Part-time faculty	93	68	132	103
Public four-year college				
Mathematics				
Full-time faculty	6,068	6,160	6,264	7,754
Part-time faculty	876	1,339	2,319	3,337
Private four-year college				
Mathematics				
Full-time faculty	3,352	3,579	4,153	4,762
Part-time faculty	945	1,359	2,099	2,706
Total, university and four-year college				
Full-time faculty	16,355	15,876	16,632	18,511
Part-time faculty	2,529	3,465	5,588	7,190
Two-year college				
Mathematics				
Full-time faculty	4,879	5,944	5,623	6,277
Part-time faculty	2,213	3,411	6,661	7,433
Total, all institutions				
Full-time faculty	21,234	21,820	22,255	24,788
Part-time faculty	4,742	6,876	12,249	14,623

SOURCE: Consolidated from Conference Board of the Mathematical Sciences (CBMS, 1987, pp. 41, 133).

TABLE A5.9 Mathematical sciences and computer science enrollments per full-time equivalent (FTE) of faculty

	1970	1975	1980	1985
Universities	79	85	96	105
Public colleges	78	87	105	100
Private colleges	71	73	90	73
Two-year colleges	104	123	134	118

SOURCE: Conference Board of the Mathematical Sciences Surveys (CBMS, 1987, p. 44).

TABLE A5.10 Age distribution of full-time mathematical sciences faculty in 1985 in four-year colleges and universities

Age	Percent of faculty	Retirement years at age 67	Estimated number of retirements[a] (per year)	Estimated number of doctorate retirements[b] (per year)
≥60	7	1986-1992	186	138
55-59	8	1993-1997	297	220
50-54	14	1998-2002	520	385
45-49	17	2003-2007	632	468
40-44	19	2008-2012	706	522
35-39	15	2013-2017	—	—
30-34	14	2018-2022	—	—
<30	6	2023-	—	—

[a] Based on a total of 18,600 (see Table A5.8).
[b] Assumes that 74 percent of retirees have doctorate.
SOURCE: Adapted from Conference Board of the Mathematical Sciences (CBMS, 1987, p. 57).

A Challenge of Numbers

TABLE A5.11 Age distribution of full-time mathematical sciences faculty in 1985 in two-year colleges

Age	Percent of faculty	Retirement years at age 67	Estimated number of retirements[a] (per year)
≥60	4	1986-1992	36
55-59	7	1993-1997	88
50-54	13	1998-2002	163
45-49	18	2003-2007	226
40-44	24	2008-2012	301
35-39	18	2013-2017	—
30-34	11	2018-2022	—
<30	5	2023 -	—

[a] Based on a total of 6,300 (see Table A5.8).
SOURCE: Adapted from Conference Board of the Mathematical Sciences (CBMS, 1987, p. 137).

TABLE A5.12 Employment status of new doctorates awarded by U.S. and Canadian mathematical sciences departments

	1977	1978	1979	1980	1981	1982	1983	1984	1985	1986	1987
Total	972	952	889	858	904	860	792	789	769	801	845
Doctoral departments	296	280	267	239	235	239	220	192	206	225	202
Master's/Bachelor's depts.	172	160	155	154	157	138	159	175	138	128	141
Other academic units	110[a]	84[a]	74	68	91	85	60	81	67	72	85
Research institutes	—	—	12	28	26	35	15	22	24	16	28
Government	62	44	34	37	28	24	24	23	14	27	19
Business and industry	136	166	168	167	169	146	105	111	108	109	104
Outside United States	159	182	145	137	172	151	171	153	187	185	201
Other	37	36	34	28	26	42	38	32	25	39	65

[a] These include research institutes.
SOURCE: American Mathematical Society (AMS, 1977 to 1987).

TABLE A5.13 Mathematical sciences faculty salaries by type of institution (in 1985 constant dollars)

	1970	1975	1980	1985
Group I				
Assistant professor	$29,400	$25,600	$25,900	$27,700
Associate professor	$36,800	$33,200	$32,800	$34,700
Full professor	$57,000	$52,300	$49,500	$48,800
Group II				
Assistant professor	$30,900	$27,100	$25,400	$27,200
Associate professor	$38,300	$33,600	$32,100	$33,200
Full professor	$54,700	$48,400	$44,200	$44,400
Group III				
Assistant professor	$30,900	$26,600	$24,500	$26,600
Associate professor	$38,500	$33,400	$31,400	$31,900
Full professor	$51,500	$44,000	$41,900	$42,400
Group IV (statistics)				
Assistant professor	$30,400	$27,000	$26,400	$28,700
Associate professor	$36,800	$35,100	$34,700	$34,600
Full professor	$49,400	$50,700	$48,100	$50,000
Group M				
Assistant professor	$29,900	$26,500	$25,100	$26,100
Associate professor	$35,500	$32,700	$30,600	$31,600
Full professor	$44,600	$39,700	$38,200	$39,200
Group B				
Assistant professor	$28,100	$23,600	$22,400	$24,800
Associate professor	$35,000	$28,600	$27,600	$29,400
Full professor	$40,800	$35,800	$33,600	$34,700

NOTE: The American Mathematical Society classifies the mathematical sciences institutions by their ranking in the 1982 assessment and by degree programs offered. Of the doctorate-granting institutions, Group I consists of the 39 top-ranked, Group II the next 43, and Group III the remaining 73. Group IV consists of the 69 departments (or programs) of statistics that offer doctoral degrees, Group M consists of 273 master's-granting institutions, and Group B consists of 950 bachelor's-granting institutions. (See Box 3.2 for more detail.)
SOURCE: American Mathematical Society (AMS, 1986).

MAY 0 7 1990